WIRING FOR WIRELESS SITES

Ira Wiesenfeld, P.E.

NEW ENGLAND INSTITUTE
OF TECHNOLOGY
LIBRARY

THOMSON
---*---
DELMAR LEARNING

Australia • Canada • Mexico • Singapore • Spain • United Kingdom • United States

Wiring for Wireless Sites
by Ira Wiesenfeld, P.E.

Business Unit Director:
Alar Elken

Executive Editor:
Sandy Clark

Senior Acquisitions Editor:
Gregory L. Clayton

Senior Development Editor:
Michelle Ruelos Cannistraci

Editorial Assistant:
Jennifer Luck

Executive Marketing Manager:
Maura Theriault

Marketing Coordinator:
Brian McGrath

Executive Production Manager:
Mary Ellen Black

Production Manager:
Andrew Crouth

COPYRIGHT © 2003 by Delmar Learning, a division of Thomson Learning, Inc. Thomson Learning' is a trademark used herein under license.

Printed in Canada

1 2 3 4 5 XX 05 04 03 02 01

For more information contact
Delmar
5 Maxwell Drive
Clifton Park, N.Y. 12065

Or find us on the World Wide Web at
http://www.delmar.com

ALL RIGHTS RESERVED.
No part of this work covered by the copyright hereon may be reproduced in any form or by any means—graphic, electronic, or mechanical, including photocopying, recording, taping, Web distribution, or information storage and retrieval systems—without the written permission of the publisher.

For permission to use material from the text or product, contact us by
Tel. (800) 730-2214
Fax (800) 730-2215
www.thomsonrights.com

Library of Congress Cataloging-in-Publication Data:

Library of Congress Control Number: 2002107523

Wiring for Wireless Sites/
Ira Wiesenfeld

ISBN: 1-4018-1037-3

NOTICE TO THE READER

Publisher does not warrant or guarantee any of the products described herein or perform any independent analysis in connection with any of the product information contained herein. Publisher does not assume, and expressly disclaims, any obligation to obtain and include information other than that provided to it by the manufacturer.

The reader is expressly warned to consider and adopt all safety precautions that might be indicated by the activities herein and to avoid all potential hazards. By following the instructions contained herein, the reader willingly assumes all risks in connection with such instructions.

The publisher makes no representation or warranties of any kind, including but not limited to, the warranties of fitness for particular purpose or merchantability, nor are any such representations implied with respect to the material set forth herein, and the publisher takes no responsibility with respect to such material. The publisher shall not be liable for any special, consequential, or exemplary damages resulting, in whole or part, from the readers' use of, or reliance upon, this material.

DEDICATION

To Natalie, my best friend and wife of 29+ years, who has stood by me in all endeavors, and whose understanding and caring has made my life complete.

Contents at a Glance

Chapter 1:   Wireless Sites 1

Chapter 2:   Equipment Layout 23

Chapter 3:   Equipment Wiring 43

Chapter 4:   AC Wiring 65

Chapter 5:   24 VDC and 48 VDC Wiring 87

Chapter 6:   RF Cabling 101

Chapter 7:   Antenna Installation 123

Chapter 8:   Telephone Wiring...163

Chapter 9:   Grounding 181

Chapter 10: Lightning Protection 195

Chapter 11: Miscellaneous Wiring and Cabling 211

Chapter 12: Emerging Technologies 227

Table of Contents

Preface xv
Author Biography xvii
Acknowledgements xviii
Introduction xix

CHAPTER 1

Wireless Sites 1
Objectives 1
Introduction 1
What Makes a Site? 2
Buildings at the Base of a Tower 4
Building Penthouse or Rooftop 5
Pads at Tower Sites 6
Radio Equipment at a Site 7
Power Requirements 8
Air Conditioning Requirements 9
Weatherproofing Requirements 10
Wiring 10
Lighting 12
Lightning Protection 13
Grounding 13
Security 15
Safety 16

Licenses Required for Installation 18
Site Planning Worksheets 18
Standards for Installation 19
Fire Protection and Plenum 19
Acceptance Test Procedure 21
Summary 21
Questions for Review 22

CHAPTER 2

EQUIPMENT LAYOUT 23

Objectives 23
Key Terms 23
Introduction 24
Safety Precaution 24
Equipment Placement 26
Cable Entrances 28
Overhead Structure 30
Raised Floor 36
Anchoring Equipment 37
Weight Loading 38
Drawings 40
Summary 40
Questions for Review 41

CHAPTER 3

EQUIPMENT WIRING 43

Objectives 43
Key Terms 43
Introduction 44
Standards 44
Inter-Bay Wiring 44

External Wiring 45
Punchdown Blocks 50
WireWrap Block 51
Data Communications Interfaces 51
Coaxial Cables 57
Alarm Wiring 57
Telephone Wiring 57
Fiber-Optic Cable 58
Cables 59
Connectors 60
Labels 60
Drawings 61
Testing the Wiring 61
Summary 62
Questions for Review 63

CHAPTER 4

AC Wiring 65
Objectives 65
Key Terms 66
Introduction 66
National Electrical Code 66
Safety 67
Licenses 68
AC Requirements 68
Backup Generators 69
Load Centers 71
Uninterruptible Power Supplies 73
Power Outlets 75
Lighting 76
Grounds and Neutral 77
Tower Lights 78

Lightning Protection 80
Testing 81
Documentation 83
Summary 83
Questions for Review 85

CHAPTER 5

24 VDC AND 48 VDC WIRING 87

Objectives 87
Key Terms 87
Introduction 88
Safety Precautions 88
Battery String 88
Charger 91
Connectors 92
Cables 92
Power Distribution Panels 93
Fuses and Circuit Breakers 93
Alarms 97
Testing 97
Documentation 98
Summary 99
Questions for Review 100

CHAPTER 6

RF CABLING 101

Objectives 101
Key Terms 101
Introduction 102
Safety Precautions 102
Cable Loss 103

Cables	103
Connectors	106
Jumpers	116
Lightning Protectors	116
Testing	117
Summary	118
Questions for Review	121

CHAPTER 7

ANTENNA INSTALLATION 123

Objectives	123
Key Terms	123
Introduction	124
Disclaimer on Installation and Climbing	124
Safety	125
Antenna Types	125
Mountings	147
Installation Techniques	149
Connector Installations	151
Installing Cables	152
Testing	156
Gas Filled Transmission Lines	160
Summary	160
Questions for Review	161

CHAPTER 8

TELEPHONE WIRING 163

Objectives	163
Key Terms	163
Introduction	164
Telephone Lines	164

Telephone Stations 166
Telephone Board 166
Main Frame 167
Wiring 168
Connectors 168
Jumpers 169
Lightning Protection 170
Private Line Circuits 172
Summary 176
Questions for Review 179

Chapter 9

Grounding 181

Objectives 181
Key Terms 181
Introduction 182
What Is a Ground? 182
Tower Ground 182
Outside Ring Ground 183
Inside Ring Ground 183
Ground Bus 183
Equipment Grounding 185
Coaxial Cable Grounding 186
Grounding Other Equipment 187
Ground Rods 188
Cables 188
Connectors 191
Cadwelds 191
Other Grounds 191
Testing 193
Summary 193
Questions for Review 194

CHAPTER 10

LIGHTNING PROTECTION 195

Objectives 195
Key Terms 195
Introduction 196
Why Do We Need Lightning Protection? 197
Cables 197
Connectors 197
AC Protection 198
Telephone Line Protection 201
Antenna Protection 202
Coaxial Cable Protection 204
RF Equipment Protection 206
Summary 208
Questions for Review 209

CHAPTER 11

MISCELLANEOUS WIRING AND CABLING 211

Objectives 211
Key Terms 212
Introduction 212
Data Circuits 213
Alarms and Sensors 213
Cables and Connectors 216
Low Voltage Wiring License 216
Cable Testing 217
Fiber Optic Equipment Requirements 217
Fiber Optic Connector Types 218
Protecting Fiber Cables 222
Fiber Optic Cleaning 222
Fiber Optic Testing 223

Summary 223
Questions for Review 225

CHAPTER 12

EMERGING TECHNOLOGIES 227
Objectives 227
Introduction 227
Networking in the Metropolitan Environment 228
Networking in the Enterprise Environment 230
Networking for the Individual 231
700 MHz Systems 233
RFID 234
Summary 234

APPENDIX A

ANSWERS TO ODD-NUMBERED QUESTIONS 235
Visit the Delmar Thomson Learning electronics technology website, www.electronictech.com, for the answers to all of the questions.

APPENDIX B

GLOSSARY 243

APPENDIX C

ADDITIONAL RESOURCES 253

INDEX 255

Preface

Twenty years ago, all telephones had cords that were plugged into a wall, plumbers and doctors were the only people who wore pagers, and police cars and fire trucks were the main users of the radio spectrum. Today, that has changed. Cellular telephones and pagers are commonplace; everyone seems to be "wireless". Being wireless requires that many technologies be wired together. These multiple technologies merge at the antenna sites and towers, where the signals become wireless.

All these developments have created a demand for technicians and engineers to work in the wireless industry. There are many books and training programs to explain how the radios and electronic circuits work. Until this book, however, there was not an authoritative reference book to explain all of the intricacies involved with the planning and installation of wireless communication sites.

This book is intended to explain what is required in the planning and implementation stages of a wireless system, identify what is required for the installation, and serve as a reference guide for maintenance of the wiring at a site. Students just getting into the business, professionals in the business, maintenance technicians, and engineers will all want to keep this book at the site for quick answers to issues that may arise.

I have been involved with radio sites since 1966, and I have always had to leave the site to find reference materials on connectors or cables, and then go back to the site to correct whatever problem I had originally gone to the site to fix. In addition, the engineers never had a good standard for the system documentation, so this book will train future engineers and technicians in what maintenance technicians need to know and have at their disposal in order to properly maintain a site.

It does not matter whether you have one day or three decades of experience; you will find this book useful every time you enter a radio site. Also, there are no prerequisites or assumed knowledge required for you to successfully understand this material. Every concept and bit of information you will need to work at a site is presented here.

The book is arranged by the types of systems found at a site. The first chapter introduces you to the overall description of a site. The later chapters are divided by the specific function of the components of a site.

I will use graphics, pictures, and illustrations to help describe the topics. I also provide a list of additional resources. Quick background information is provided in the "Concept for Review" boxed articles. Use these aids to better prepare yourself as a professional.

This book is designed to be a reference guide. That is, the chapters allow you to quickly locate the information you are seeking. The sub-headings are clear and logical to further steer you to exactly the bit of information you need.

I've included questions to ensure that you understand the main concepts from each chapter. Answers to the odd-numbered questions are found at the back of the book in the appendix. You can find the answers to all of the questions online, at the Delmar Thomson Learning electronics technology website, www.electronictech.com.

Author Biography

The author, Ira Wiesenfeld, P.E. has been in the commercial radio business for a long time and has built many radio systems and sites over the last three and a half decades. He has a BSEE from SMU in Dallas, Texas, is a licensed professional engineer in the state of Texas, has been involved with amateur radio since 1963, and holds a commercial FCC General Radiotelephone Operator License.

Ira can be reached via email at iwiesenfel@aol.com.

Acknowledgements

I want to thank Dennis Riise, Bill Caldwell, Jr., and Robert H. Smith for the content they contributed, and Bill Tull for reviewing the book. I also wish to thank Greg Clayton, Michelle Ruelos Cannistraci, Debbie Abshier, Joell Smith-Borne, Phil Velikan, Debbie Berman, Linda Malcak, Kim Heusel, and the entire project team at Abshier House and Delmar Learning for their assistance in getting this work from our computers into this book.

Ira Wiesenfeld, P.E.

Introduction

There are many books in existence today that adequately describe how radio and wireless equipment works. What does not appear to be in existence is a comprehensive book that explains what you need to plan, install, and maintain wireless communication sites. This book is written for the professional who has to install or maintain a wireless radio site. It is also written to inform those not already in the wireless business that there are many items that must be taken into consideration to create a site plan and perform an installation.

With the proliferation of cellular, trunking, paging, PCS, wireless data, and all of the other services that use wireless technologies to communicate, the need for radio channels and sites has grown tremendously in the last twenty years. As the number and size of sites have grown, so have the problems associated with poor site planning and installations. We hope that this book answers many questions about radio sites.

Racks, wires, and cables do not need maintenance typically. If you have a problem with the installation, wiring, or cabling part of a system that is not caused by the components, connectors, or something external, then the installation was not done properly. The physical installation should last practically forever. Of course, preventive maintenance must be done in some situations to prevent wiring damage due to weather, chafing, or other factors.

This book will explain some of the fundamentals of installation, wiring, and cabling, as well as the details of installing common wires and cabling found in a wireless radio system.

Radio Tower

This book is not a novel or coffee table book, nor is it meant to be a bedside reading book. This book has been written for two purposes. The first is to teach technicians and engineers how to wire a wireless radio site. The second purpose is to provide a reference for those in the business. Keep this reference "alive" with your own notes and comments. Some carriers and site mangers have their own standards and rules, and you must follow those requirements at their sites. If you keep notes about those requirements in this book, you will always have everything right where you need it.

CHAPTER 1

WIRELESS SITES

OBJECTIVES

After completing this chapter, you will understand the following concepts:

- The requirements for planning a wireless site
- The types of commonly used wireless sites
- The components found at the site
- The standards requirements that are applied to wireless sites

INTRODUCTION

A radio system usually consists of the hardware for the system, and, being wireless, an antenna. In order to provide adequate range for the radio system, the antenna is usually located fairly high in relation to the ground. This can be accomplished by mounting it on a pole, tower, or the top of a tall building. In fact, all of these types of facilities are used in radio sites. Besides the physical building or part of a building, or pad, there are many items that must be present to make the area into a complete radio site. These include the following:

- Physical space
- Heating or air conditioning

- Equipment
- Power
- Backup power
- Telephone lines
- Wiring
- Fire protection
- Planning
- Documentation
- Safety
- Commissioning

What starts out as a simple system requires an enormous amount of peripheral equipment to create an acceptable radio site. See figure 1.1 and figure 1.2 for a look at a typical radio site, inside and out.

FIGURE 1.1 Radio site exterior

CHAPTER 1: WIRELESS SITES

FIGURE 1.2   Radio site interior

WHAT MAKES A SITE?

Why do you need all those things? The most basic requirements for a radio site are the radio equipment, antenna support, and the antenna(s) (see figure 1.3). Because the radio equipment needs power, though, you need a power system. Since the AC power can and does fail periodically, you need backup power too. This usually means a battery power system. All of this equipment generates heat, so you need an air conditioning system. The antenna is usually one of the highest points in the vicinity, so you have to add lightning protection. Lightning protection needs a good ground, so you need a ground system. All of this equipment can catch fire, especially due to lightning, so you have to have a fire protection system. The equipment needs periodic maintenance, so you need AC power and lighting for the personnel who visit the site. Besides all of this, many sites are built with the hopes that the site will grow (and provide a corresponding increase of revenue), so you need to build reserve capacity into every one of the components mentioned.

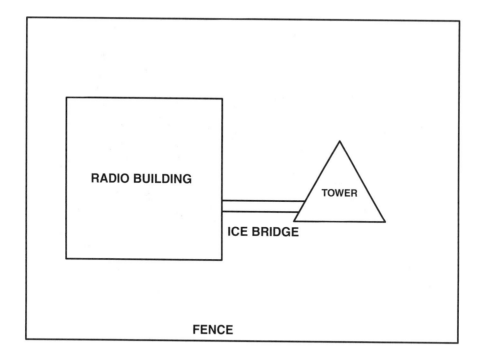

FIGURE 1.3   A diagram of a typical site

Buildings at the Base of a Tower

Many sites are small buildings that are located right next to radio towers (see figure 1.4). Many of the buildings in existence today are prefab buildings that have been built especially for radio sites. There are other buildings that are built on-site and contain the radios. Some site owners have purchased railroad boxcars and made them into radio sites. Others have used semi-truck trailers for this. The size of the building varies from just a few square feet to several thousand square feet. One of the disadvantages to this type of setup is that the distance you'll need to run the cables from the antennas to the building is longer than with the other setups. Longer cable runs mean more signal loss and greater costs.

The amount of equipment dictates the size of building you will need and the amount of peripheral equipment you will have to put in that building. Also, how much and how fast you plan to grow the site will determine the overall size of the building.

CHAPTER 1: WIRELESS SITES

FIGURE 1.4 A building at the base of a tower

Building Penthouse or Rooftop

In every urban area of the world, the penthouses of some of the taller buildings have been built or converted for use as radio sites (see figure 1.5). The same factors that are present in stand-alone buildings are also found in these sites. The radio system operators and engineers like these sites best, because the Radio Frequency line loss is minimized thanks to the shorter runs of coaxial cables that are found on rooftop locations. The downside is the high rent that you'll have to pay for these sites. Rooftop leases are expensive, often costing many dollars per square inch each month, while normal rental for office space is considerably cheaper, renting for only a few dollars per square foot monthly. The roof makes more money than any other part of the building.

Many penthouse sites require quite a few modifications to make a good site, including air conditioning, power, backup power, and fire protection. Some

FIGURE 1.5 Rooftop site

of the sites in these tall buildings are built with the top floor dedicated for radio and wireless systems. These dedicated sites already have all of the provisions necessary to accommodate most radio systems.

PADS AT TOWER SITES

The advent of cellular and PCS systems, and the portable handheld units that run on these systems, has made towers more common everywhere. The equipment manufacturers have met the challenge of reducing the size and power requirements of the latest equipment to allow an entire system to be placed on a concrete pad located at the base of a tower (see figure 1.6). The complete system can be housed in a single outdoor cabinet. Just add power,

FIGURE 1.6 A pad site

antennas, and telephone lines, and the system is ready to operate. The disadvantage to this system is that these small cabinet systems are a little bit more vulnerable to severe weather than other types of sites. People who will be working at the site will be exposed to the weather as well.

RADIO EQUIPMENT AT A SITE

The radio equipment that you will need at the site can be as simple as a single radio base station or transmitter connected to a coaxial transmission line and to a single antenna. Or it may be a complete cellular site with many bays of equipment, a Fiber-Optic interface terminal, a dozen or so antennas, and a huge power supply system. You may also require many support peripherals such as computers, printers, CRT terminals, modems, telephone lines, monitoring systems, and other items. Systems from different manufacturers require different peripherals.

FIGURE 1.7 Load center

POWER REQUIREMENTS

All electronic equipment requires power. Some systems use 120 VAC or 240 VAC, while others use 24 VDC or 48 VDC. The total current consumption of the site determines the class of service for the site. The load center is the breaker box that each outlet, lighting circuit, and piece of equipment goes to (see figure 1.7). The 24 VDC or 48 VDC system has a string of batteries and a power charger, and this uses either 120 VAC or 240 VAC. In addition

FIGURE 1.8 Generator

to the main power from the commercial power vendor, the site usually has a backup power source such as a generator, fuel cell, or solar power panel (see figure 1.8).

AIR CONDITIONING REQUIREMENTS

Because all equipment generates heat, almost every site has an air conditioning system. Regardless of whether it is hot or cold outside, the site requires interior cooling. Most manufacturers of cellular and PCS systems require redundant air conditioning systems (see figure 1.9). A small site has a small air conditioning system. A large site has multiple redundant systems. The more equipment inside the building, the more air conditioning required.

FIGURE 1.9 Redundant air conditioning systems

WEATHERPROOFING REQUIREMENTS

Most radio systems have the equipment indoors or in weatherproof cabinets. Almost every system has the antennas outdoors on a tower or building rooftop. Regardless of whether the equipment is housed in a building or a cabinet, the facility must be weatherproofed to keep the rain, snow, dirt, and moisture away from the equipment. This includes all doors, cable entrances, exterior cable connections, and any other transition from the inside to the outside (see figure 1.10). Poor weatherproofing leads to corrosion and poor system performance.

WIRING

A site has wiring for the following purposes:

- AC Power
- DC Power

CHAPTER 1: WIRELESS SITES

FIGURE 1.10 Cable boot entrance panel

- Antenna(s)
- Inter-bay wiring
- Alarms and sensors
- Interconnection
- Telephones
- Data circuits
- Controllers

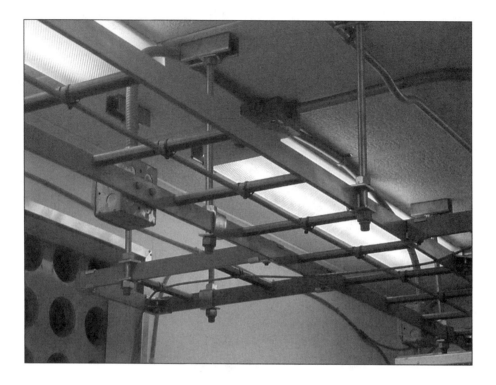

FIGURE 1.11 Cable rack at a site

The way this wiring is installed is very important, and there are many standards and techniques that must be observed in the initial construction of the site and system (see figure 1.11). These will be covered in much greater detail later in this book.

LIGHTING

In the development of many sites, the lighting is left up to the system. The majority of the sites use fluorescent lights for many reasons, including the low initial and operating costs, the length of time before you have to replace bulbs, and light output, as well as (often) because that's the way they did it in the rest of a building.

Lightning Protection

Because a radio site has an antenna that is mounted high above ground, it is the natural target of lightning strikes. You'll need to consider many factors regarding lightning. Chapter 10 is dedicated to lightning protection. In addition to entering via the antenna transmission line, lightning can enter a site or system via the AC power lines and telephone or data lines. By providing the proper protection devices, you can limit the amount of damage to a site from a lightning strike from most hits. A full direct hit will still destroy much of a site. Fortunately, most lightning hits are not full direct strikes.

Grounding

Once lightning does enter a site, you can reduce or even eliminate the damage if every item in the site is at the same potential. This is the purpose of grounding (see figure 1.12).

To accomplish this, tie together the following items using a very heavy gauge of wire, usually #8 or bigger, and then terminate this wire into the single ground rod that is located near the tower support:

- Tower
- Coaxial transmission line
- AC load center
- AC generator
- UPS
- Equipment racks
- Lightning protectors
- Metal storage cabinets
- Metal desks

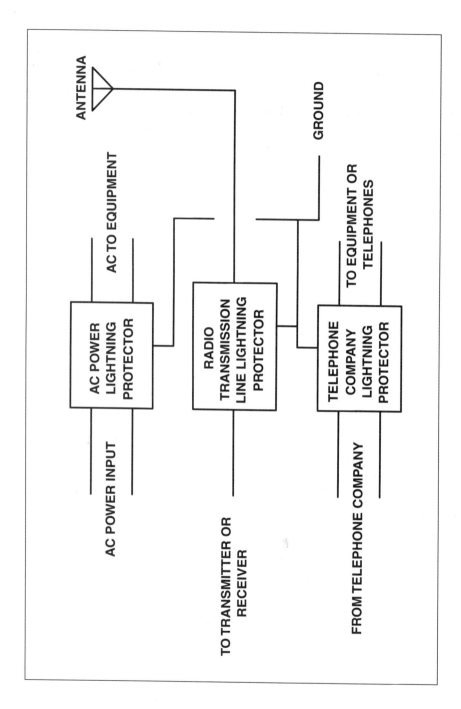

FIGURE 1.12 Lightning protection devices and grounding

CHAPTER 1: WIRELESS SITES

FIGURE 1.13   Ground lug on equipment

- Metal door frames
- Air conditioning equipment frames

It is not sufficient to have the wires connected to every item at the site; the connections themselves must be able to withstand the high current of a lightning strike. This means that special tools must be used to connect the wires to the lugs, and the lugs must be properly attached to the various items (see figure 1.13).

SECURITY

Radio sites are usually in unattended buildings in remote areas. Because of this, they are vulnerable to theft and vandalism. Most sites have alarms that tie back to a central system that notifies an operator that a site has been

FIGURE 1.14 An alarm panel by the door of a site

entered (see figure 1.14). The security system also monitors other parameters of a site, such as temperature, power outage, smoke, or fire.

Many sites have multiple tenants. In these cases, access to the sites requires multiple locks (see figure 1.15). Be sure to not disable access for other tenants if you replace the locks.

SAFETY

The FCC Office of Technology has done extensive work on the biological effects of Radio Frequency energy. RF levels above a certain intensity can cause permanent damage to humans. All sites that exceed a specific power level must have a maximum permissible exposure study, and RF safety signs must be posted (see figure 1.16).

CHAPTER 1: WIRELESS SITES

FIGURE 1.15 Series linked locks

FIGURE 1.16
An RF safety sign at a site

The AC and DC power sources have the potential to be lethal to humans. Because of the dangers here, only skilled persons should be accessing the wiring associated with the power sources. Most local government jurisdictions have rules and regulations regarding the installations of these items.

All personnel entering a site should be aware of the hazards at a site.

LICENSES REQUIRED FOR INSTALLATION

Most sites require several licenses and registrations with numerous governmental agencies to get established. These include the following:

- Zoning board permits
- Building permits
- FCC licenses
- FAA permits
- FCC tower registration
- Utility inspections

The process for obtaining these permits and licenses is beyond the scope of this book and course. Do not start construction on a site without these permits in hand, though, since the penalty for doing so is usually an immediate rejection from ever receiving a license. The company or organization that holds these licenses is the entity that is ultimately responsible for the site. Installers and other technicians may have contractual obligations, but the licensee has final responsibility for everything that happens at the site.

SITE PLANNING WORKSHEETS

Most manufacturers have planning worksheets for putting together a new site or for making an addition to a site. If you are adding a radio to a site or are not associated with one of these companies, you are on your own with the planning. It is strongly advised that you make up your own worksheet in these cases, so that you do not skip any item that is needed for a proper site installation.

STANDARDS FOR INSTALLATION

A variety of organizations provide specifications for proper site installation. Many of these specifications are not just suggestions, but the law! Check with local authorities during the site planning stage about these requirements. Several standards organizations are listed here:

- NEC – National Electric Code
- NFPA – National Fire Protection Agency
- UL – Underwriter's Laboratories

Most of the manufacturers have standards that must be followed closely during the installation phase of a project. These include power source requirements, grounding, floor space, temperature concerns, and connections. Be sure to follow all of these standards.

FIRE PROTECTION AND PLENUM

Most municipal jurisdictions require some type of fire protection at all building structures. Since water is dangerous and inappropriate for electrical fires, an alternate fire suppression system is required. A substance called Halon was the most popular material used for fire suppression (see figure 1.17). It works by blanketing the fire and starving the fire of oxygen. People cannot be in a room that has Halon released in it, though, and it is not ozone friendly. New installations are using FE36 now, which is less damaging to the ozone. The tanks and release systems are the same, only the chemical released is different (see figure 1.18).

Some buildings and localities have requirements concerning the toxicity of the products of combustion from burned materials during a fire. These

FIGURE 1.17 Halon alarm bell

FIGURE 1.18
Halon storage tank

buildings require that all toxic wires be run through in metal conduits, and that any exposed wires have a plenum rating. The cable industry has responded by making cables with plenum rated sheathing. Plenum cable uses nontoxic materials in the insulation of the cable and costs substantially more than regular cable.

Acceptance Test Procedure

When a site is completed, a formal Acceptance Test Procedure should be performed. This confirms that all parts of a system have been completed and that someone has taken responsibility for the installation.

Documentation

It is important to document everything at your site. Documentation is important to prove that you have the proper permits and licenses to construct and operate your site. Documenting all aspects of the wiring and power systems will also reduce the amount of time required to troubleshoot a site or conduct preventive maintenance.

Summary

From this chapter, you have learned that the installation of a radio system involves much more than adding a radio and an antenna to a building. A radio system can be placed in several locations, such as on a rooftop, in a building at the base of a tower, or in a weatherproof cabinet on the tower's concrete pad. In each case, you must take the power requirements, fire protection, and security of the site into consideration. All sites require lightning protection, which means you must install a grounding system. The equipment is expensive, so you'll need security as well. Licenses and permits are mandatory in many instances, and you'll have to obtain those before a site is even constructed. You should also document the completion of any site or system. It is a complicated process, but fortunately, a good installation will last for years.

QUESTIONS FOR REVIEW

1. Name at least three things you need to take into consideration to ensure a safe site.

2. Name the items typically found at a site.

3. What are the three types of sites?

4. What are the advantages and disadvantages of the various site types?

5. Name three specification agencies that specify requirements at sites.

6. Why is lightning protection important in areas where lightning does not strike often?

7. What is a ground?

8. What types of components must be grounded?

9. What types of cabling and wiring are found at a site?

10. What is the result of inadequate weatherproofing?

11. Who is responsible for a site?

12. Name three items you will be required to have to receive a site permit or license.

13. What is Halon?

14. Why is plenum grade wiring a requirement in some localities?

15. Is air conditioning required at a site during cold weather?

16. Why is battery backup required?

17. Why is site documentation important?

CHAPTER 2

EQUIPMENT LAYOUT

OBJECTIVES

After completing this chapter, you will understand the following concepts:

- Safety precautions that must be observed with respect to the equipment
- Placement of equipment at a site
- Proper cable placement and protection for cable ingress and egress (going in and out of the building)
- Considerations for overhead structures and raised floors
- Methods of anchoring the equipment in place
- The implications of weight loading
- The relation between drawings and equipment placement

KEY TERMS

- Multi-bay equipment
- Plumb
- Overhead rack
- Raised floor

- Anchors
- Stud guns
- Floor loading

INTRODUCTION

All radio sites have a layout plan for how the equipment integrates into the available space. This chapter explains the parameters that are part of the physical installation of the equipment at a wireless site. Whether you are planning a small site or a large site, you still have these same parameters to consider (see figure 2.1).

FIGURE 2.1 The interior of a radio site

SAFETY PRECAUTION

A radio site is filled with hazards that will not be immediately obvious to people not familiar with a site. Some of these hazards are immediate, and some are potential hazards. This chapter will only concern itself with those hazards related to the equipment at the site.

CHAPTER 2: EQUIPMENT LAYOUT

Immediate hazards include the following:

- Equipment falling on the personnel constructing the site
- Sharp edges of metal pieces on which people could be hurt
- Improper or unsafe use of the construction tools
- Floors that cannot withstand the weight of the equipment
- Obstructions in walkways

CONCEPT FOR REVIEW:

EMERGENCY RESPONSE PLAN

Each site should have a well developed plan for dealing with emergencies. OHSA requires this plan to be put in place **before** construction begins. Review the emergency response plan at the first planning session to ensure compliance. All members of the construction and installation team should be well versed in procedures to follow in case of an emergency. Each member should have basic first aid training, and key team members should have advanced medical training to provide aid until emergency response personnel arrive. If the site is located in a remote area, notify local emergency personnel that you are going to be out there working, and provide them with your location. Here are a few suggestions for things the site emergency response plan should include:

- A list of emergency phone numbers
- The location of communications equipment
- Well-marked locations for first aid supplies and rescue equipment
- Clearly marked exits
- Well-identified temporary danger zones
- Clearly identified hazardous materials and the actions to take in case of accident

FIGURE 2.2 Planned site expansion

The safety of the construction personnel and operating personnel should always be paramount in both the construction phase of a project and in the completed project. At no time should safety ever be compromised. Some of the manufacturers have standards that insure these safety measures, while others do not. Always work to the highest standards available.

Equipment Placement

The exact placement of the equipment at a site is usually determined by the systems engineer of the operating company or by the site manager of shared sites. In these cases, the exact placement of the equipment is predetermined before the installation team arrives on-site (see figure 2.2).

On multi-bay equipment systems, the placement of the equipment racks is dictated by the inter-bay cable lengths. Most manufacturers have premade

FIGURE 2.3 A level placed on an equipment rack frame

these cables at their factory, and this will limit how the bays are put together at the installation site. The factory will provide drawings that show proper equipment and cable arrangement. On single bay systems, there is no restriction on the placement of the bay. It just needs to integrate into the lineup of the rest of the site. Most sites have a site manager that will determine the placement of all equipment at the site. Do not install equipment without placement approval by the site manager.

It is important for both aesthetic reasons and long-term support that all equipment is level, or plumb, in all dimensions. A good carpenter's level should be used to ensure that the equipment is level in all three planes (see figure 2.3). The typical industry tolerance is usually for the level indicator

FIGURE 2.4   Reading the level

bubble to just touch the indicator line on the level (see figure 2.4). There is *no* excuse for this not being exact. Non-level equipment racks or stands will cause problems later in the installation process. Do it right the first time!

You may need to consider other site layout issues, such as fire exits, distance between rows of equipment cabinets, and the accommodations specified by the Americans with Disabilities Act (ADA).

CABLE ENTRANCES

All radio sites will have cables running between the inside of the site and the outside. There is always the antenna, which is mounted outside. There may also be power lines, telephone lines, and grounding cables, depending upon the site.

There are a number of good reasons for making sure that all your cable entrances are secure. Because of environmental issues, they must be weatherproofed. You must also weatherproof exterior cable connections. In addition,

FIGURE 2.5 The transmission line cable entrance plate

the site must be protected against varmints. These include rodents, birds, bees, wasps, snakes, spiders, scorpions, centipedes, millipedes, and the like. Communication sites are typically warm and comfortable compared with the surrounding environment. All entrances to the site must be plugged or filled (see figure 2.5 and figure 2.6). Rodents will eat through expansive foam, so use steel wool in irregular openings. Holes and conduit entrances should be capped with metal or PVC caps when possible. Nature's creatures have destroyed many pieces of equipment. They can also affect maintenance personnel.

Besides needing to be protected from weather and bugs, many sites have additional requirements that relate to the spread of fire or smoke in a building. Once again, check with local ordinances concerning room to room and floor to floor wiring chases. All of these factors must be observed when installing the cables.

FIGURE 2.6 A grounding cable through a wall with seal

The cables should be cut to length, so that there is no excess cable or slack in the building. Be sure to consider the locations of the doors of the building as well as the cable entrances as you decide where to place the equipment.

OVERHEAD STRUCTURE

Most sites use a series of overhead racks to hold and secure the inter-bay cables and the external cables. In addition, these racks are secured to both the walls and the equipment bays, and this gives the bays the necessary support to prevent them from falling over. In earthquake zones, there are special requirements for added support. The overhead racks can be supported by the equipment, by mounts on the walls (see figure 2.7), by supports coming from the ceiling (see figure 2.8), and in some locations, by vertical extensions from the floor.

FIGURE 2.7 A rack tied to a wall

FIGURE 2.8 A vertical support tied to the ceiling

CONCEPT FOR REVIEW:

CLASSIFICATIONS OF FIRE EXTINGUISHERS

There are four types of classifications of portable fire extinguishers. These classifications are designated by the types of burning materials the fire extinguisher agent is optimized to extinguish. The extinguishing agents can be easily remembered by mnemonics regarding the way the designated burning material would act if you used water on it. Note that water is **not** the extinguishing agent of choice for many materials!

Class A — Ordinary Combustibles

This group includes most of the common flammable materials.

- Paper
- Wood
- Cloth
- Plastics and rubber

Action of water: Typically extinguishes this type of fire. Dry chemicals are often used on small, contained areas.

Mnemonic: A O K (Class **A** **O**rdinary **K**umbustibles)

Class B — Combustible Liquids

This type of fire is extremely dangerous! Flammable liquids often have flammable vapors that can lead to severe explosions.

- Petroleum products
- Solvents
- Alcohols
- Oils
- Oil-based paints
- Lacquers

(Continued.)

(Continued.)

Action of water: Burning material typically floats on water, so water does little if anything to extinguish flame. Foam agents with densities lighter than oils may be used to float on top of the burning material and extinguish the flame.

Mnemonic: Burns. Flame continues to burn with water.

Class C — Electrical Equipment

This type of fire is a major concern to the communications facility. The key factor is that the extinguishing agent be non-conductive. Electrical components may still be energized even during the fire, and conductive extinguishing agents increase the risk to personnel.

Action of water: Water is conductive and may actually cause additional sparking, setting new fires and putting personnel at risk. Dry extinguishing agents are typically used for Class C.

Mnemonic: Conductive. Water is conductive.

Class D — Combustible Metals

This type of fire is extremely dangerous! Combustible metals produce extremely high temperatures which can melt other metals and burn through a floor, wall, or roof very rapidly.

- Sodium
- Lithium
- Potassium
- Magnesium

Action of water: Water poured on these metals may actually cause the metal to burn hotter or faster! Avoid using water on combustible metals.

Mnemonic: Detonate! Water may actually cause these materials to explode.

Check local regulations for the size (in volume) and number of portable fire extinguishers that will be required for your site. Most local fire departments offer short courses in the proper use of fire extinguishers and fire safety.

Some sites require certain bays to have an isolated ground. In these instances, special isolated ground mounting hardware is provided.

There are specially made "J" bolts or "U" bolts that are used to secure the equipment bays to the overhead rack. There are a few companies that specialize in hardware used in overhead racks, including the racks themselves. The racks are available in brass, aluminum, and steel. The site manager or system engineer should normally specify the material, and even the paint color, if used.

The overhead racks are normally ten feet in length, and available in widths from two inches up to two feet. The site engineer should specify the width required for these racks, but you will need to cut the racks down to the exact length required. The side stringers are available in one and a half inches and two inches. The site manager should also specify this dimension. The material can be hollow or solid (see figure 2.9). Most sites allow hollow rack stringers, but a few require solid stringers. The solid type allows for a much greater weight loading.

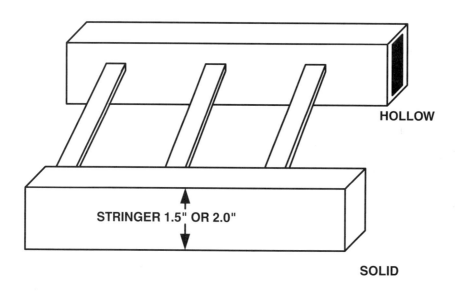

FIGURE 2.9 Cable rack types

When a rack is cut, good installation practices require that a plastic or rubber bumper be placed over the sharp edges of the cut (see figure 2.10). In fact, good installation practices will dictate that bumpers be placed on the stringers whether the rack was cut or not (see figure 2.11). Some companies require that the ends be painted to prevent rust (and to look nicer) before the bumper is placed on the stringer.

Be very careful not to spread the metal filings when cutting the racks. Protect any equipment from these filings using plastic sheets, and vacuum the floor when you are done cutting. If you do the cutting outside, be sure to sweep any filings off of the concrete, as they will leave rust stains. Metal filings always seem to end up in the wrong places. The conductive nature of metal filings poses particular risks in the communications site. Dispose of filings properly to protect your equipment and your staff.

FIGURE 2.10 A rubber bumper on an overhead rack stringer

FIGURE 2.11 An overhead rack stringer without a rubber bumper

It is important that you mount the overhead rack high enough so that people of normal height do not hit the overhead with their heads. The standard at most sites is between seven and nine feet. A measuring tape and level should be used to insure that the rack is level and has a constant height throughout the building.

RAISED FLOOR

A few sites use raised floors to allow all of the cables to be hidden beneath the floor (see figure 2.12). These sites are usually the most particular about all aspects of the installation, and almost always have an engineer who very closely monitors all aspects of the physical installation. Also, the equipment in self-contained bays must have the bottom of the cabinets cut open to enable you to run the cables down into the floors. Sometimes this has already been done, but if not, you will need to do it.

CHAPTER 2: EQUIPMENT LAYOUT

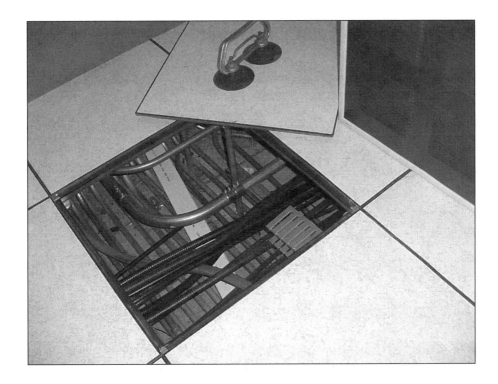

FIGURE 2.12 A raised floor site

ANCHORING EQUIPMENT

Almost all racks and bays of equipment have four mounting holes at the bottom to secure these bays to the floor (see figure 2.13). Some of the bays also have additional hardware to aid in the leveling process of the installation. Other bays and racks do not. The bays without leveling screws require shims for leveling.

The anchors come in two main styles. The first is where a lug is secured into a hole in the floor. The second is a threaded stud that is secured to the floor. The bay is placed over the stud that protrudes through the mounting hole of the equipment. Care must be taken to not destroy or alter the threads of the studs as the equipment is being installed.

37

FIGURE 2.13 Mounting hardware on a bay

You can mount the bays with standard drills or with guns that shoot studs instead of bullets. If you use a stud gun, be extremely careful and follow all safety precautions. They are powerful enough to split floor joists if not use properly—just imagine what one could do to a foot!

Whether you use anchor lugs or studs, use caution when drilling or shooting into a floor to insure that you do not damage the floor. Do not use the stud gun if you are not sure that the floor can accommodate it. These guns have destroyed many floors. In addition, in a penthouse-style installation, if the stud goes through the floor it can injure people on the floor below.

WEIGHT LOADING

All equipment has weight that you must take into consideration. Likewise, all floors have a specification that states the maximum weight loading for that given floor. It is important that the equipment weight not exceed the floor specification. Failure to observe these requirements can mean that the equipment will tilt as the floor begins to sag, or even fall through the floor at some point. Always find out the floor specification before installing the equipment.

CHAPTER 2: EQUIPMENT LAYOUT

CONCEPT FOR REVIEW:

FLOOR WEIGHT LOADING

All floors have a maximum weight load specification. It is very important to *never* exceed this value. If a cabinet has a weight that does exceed this value, take steps to distribute the weight over a wider area, and this will allow you to still use the cabinet.

Example

In this example, the floor has a maximum weight load of 900 pounds per square foot. If our rack weighs 1500 pounds, and has a base that is one and a half feet by one foot, the area is one and a half square feet, and a weight of 1500 pounds. This results in a force of 1000 pounds per square foot, exceeding the 900 pound floor weight loading specification. By putting a base on a rack that measures two feet by two feet, the area becomes four square feet, and the weight load force becomes 375 pounds per square foot, which easily allows the equipment to be placed on that floor.

Problem

> 1500 LBS
> 1.5' x 1'

1.5 × 1 = 1.5 square feet

1500 lbs/1.5 square feet = 1000 lbs/square foot

Solution

> 1500 LBS
> 2' x 2'

2 × 2 = 4 square feet

1500 lbs/4 square feet = 375 lbs/square foot

The specification for the floor is in pounds per square foot, but the specification for the equipment is by total weight. You will need to calculate the pounds per square foot for each piece of equipment by taking the weight and dividing it by the area of the floor mounting pads of the equipment. Vertical racks can hold enormous loads. Be sure that your weight loading is planned and distributed properly.

DRAWINGS

Floor placement drawings are usually provided by the equipment companies or by the end client companies. In the instances where the drawings are not provided, the installer should make very accurate drawings for future reference. This documentation also ensures that maintenance personnel have all of the proper drawings for a site. In addition, accurate equipment drawings will enable an engineer to expand or add equipment to a site.

Besides the layout drawing, there should also be a grounding drawing, an electrical power drawing, and an RF cable drawing. Don't forget the antenna and external cables, wiring, and equipment in your site drawings.

SUMMARY

Equipment is placed into a site following an orderly scheme controlled by the manufacturer, site manager, or operating company. You must perform the installation in a safe and careful manner. The equipment is anchored at the base and, in many instances, via an overhead rack system. The cables normally come out of the top of the bays, but some sites use raised floors to hide the cables underneath the floor. The site should have drawings before you start an installation, but even if these drawings are missing beforehand, they need to be provided after the installation is completed.

QUESTIONS FOR REVIEW

1. Name at least three hazards that can exist at a radio site.
2. To what standard should you work?
3. What determines the layout of a multi-bay equipment system?
4. What does "plumb" mean?
5. How close to level are you required to make the equipment?
6. Name three methods that you can use to support the overhead rack systems.
7. What precautions should you exercise when cutting racks?
8. Where do the cables come out from each equipment bay when a site has raised floors?
9. What precautions must you use if stud guns are used for anchoring equipment?
10. The specification for the floor loading is usually in what units?
11. What information is contained in the site drawings?

CHAPTER 3

EQUIPMENT WIRING

OBJECTIVES

After completing this chapter, you will understand the following concepts:

- Most of the aspects of the wiring required at a radio site
- Common methods of equipment interconnection
- Different wiring standards
- Methods for testing the cables

KEY TERMS

- WireWrap
- RJ21
- RJ11
- RJ45
- Ribbon cable
- Coaxial cable
- Fiber-Optic cable
- DB25

Wiring for Wireless Sites

Introduction

In order for any equipment to be useful at a radio site, it must be connected via wires, cables, or wireless means to other equipment. The other pieces of the system will always include an antenna, but may also include the power source and other peripheral equipment, and usually includes connectivity back to a main dispatch or controller site. This chapter will introduce some of the cabling required at a site. Later chapters will be much more specific as to the interconnections and cables required for different systems.

Standards

Site cable requirements typically specify standards that must be followed in the material and installation of a site. These standards include the following:

- Manufacturer specifications
- National Electrical Code
- Municipal codes
- Common workmanship standards for installation
- Manufacturer standards

Always work to the highest standard if more than one applies.

Inter-Bay Wiring

The inter-bay wiring is the cabling that runs between equipment racks from one manufacturer. These cables are almost always included with the equipment and are usually labeled as to where they terminate in each bay. Occasionally, you will encounter a manufacturer who has poor documentation on their inter-bay cabling, but this is usually the exception rather than the rule.

Some connectors have special markings, fixtures, or keys to show the correct insertion orientation.

CHAPTER 3: EQUIPMENT WIRING

EXTERNAL WIRING

The external wiring is almost always completed on-site, and most of the time the installers must put on the cable terminations, because they have to cut all the cables to the exact length for the particular site. This makes installing cables a much harder job on the installers.

The particular type of cable and connector varies by manufacturer and by the responsible system engineer. In this section we are going to introduce the reader to the various items that will be interconnected via the external wiring.

Some of the external wiring that must be added to complete a site installation includes the following:

- AC power
- DC power
- Voice Frequency cables
- Data communications interface cables
- RF coaxial cables
- Alarm cables
- Telephone cables
- Fiber-Optic cables

The AC power will have multiple levels of wiring. Usually the building contractor, or an electrician hired by the building contractor, is responsible for running the wiring from the AC power meter to the load center circuit breaker box (see figure 3.1). If you are not a licensed journeyman electrician, do not even attempt this part of the job. The AC wiring from the load center must be brought through conduit to the location where the wiring is to be installed.

The size of the cable must accommodate the current draw of the equipment. Besides powering the desired equipment, the AC power also has to be brought to the lighting for the site and the peripheral outlets at the site. In

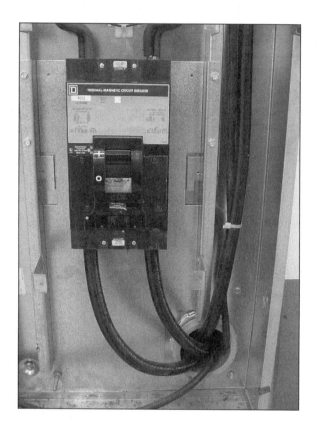

FIGURE 3.1 A close-up of the load center connection

the majority of the sites, the cable used for AC wiring is solid #16 gauge and no connectors are utilized at the terminations on either end of the wire. The wire is held by a captive screw that connects the cable to the plug or to the circuit breaker in the load center.

The DC wiring is usually in two levels also. Heavy wires with special, low resistance connectors usually come from the power board to the fuse panel of a rack or equipment bay. The current draw and the line length will determine the exact size of the cable that is required.

In bays where the equipment is powered by DC voltage, the individual shelves will have a smaller gauge wire that runs from the shelf to a power

> **CONCEPT FOR REVIEW:**
>
> # IR LOSS
>
> Ohm's law is a principle of electricity that states that the product of **CURRENT X RESISTANCE = VOLTAGE**. This is stated by the formula below:
>
> $$E = I \times R$$
>
> ### Example 1
>
> If a circuit is drawing 5 Amperes, and the resistance is 1 ohm, the voltage drop will be 5 volts.
>
> $$5 \times 1 = 5$$
>
> ### Example 2
>
> If the circuit is drawing 5 Amperes, and the resistance is 0.2 ohms, the voltage drop will be 1 volt.
>
> $$5 \times 0.2 = 1$$
>
> ### Example 3
>
> If a circuit is drawing 20 Amperes, and the resistance is 0.05 ohms, the voltage drop will be 1.0 volt. Here, we can see that even the slightest amount of resistance can cause a significant voltage drop if the current level is high.
>
> $$20 \times 0.05 = 1$$
>
> *Note: Most power systems only allow for 1 volt total loss due to IR drop.*

distribution panel, which is normally the uppermost rack in a bay (see figure 3.2).

The voice frequency cable, which is used for a myriad of different purposes, is usually a twenty-five pair or fifty pair cable. Most manufacturers use a twenty-five pair connector that is manufactured by Amphenol Corporation,

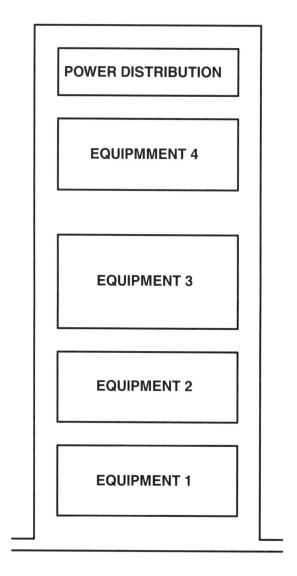

FIGURE 3.2 A typical bay

and this is called an Amphenol connector. Some companies also use the term Champ connector.

The cable normally uses the standard telephone company (telco) color code for the wires. If you are in the installation business, it would be a really good idea to memorize this code. A pair of wires will have one lead designated as

CHAPTER 3: EQUIPMENT WIRING

CONCEPT FOR REVIEW:

TELCO WIRING COLOR CODE

Pair	Ring		Tip	
1	Blue	White	White	Blue
2	Orange	White	White	Orange
3	Green	White	White	Green
4	Brown	White	White	Brown
5	Slate	White	White	Slate
6	Blue	Red	Red	Blue
7	Orange	Red	Red	Orange
8	Green	Red	Red	Green
9	Brown	Red	Red	Brown
10	Slate	Red	Red	Slate
11	Blue	Black	Black	Blue
12	Orange	Black	Black	Orange
13	Green	Black	Black	Green
14	Brown	Black	Black	Brown
15	Slate	Black	Black	Slate
16	Blue	Yellow	Yellow	Blue
17	Orange	Yellow	Yellow	Orange
18	Green	Yellow	Yellow	Green
19	Brown	Yellow	Yellow	Brown
20	Slate	Yellow	Yellow	Slate
21	Blue	Violet	Violet	Blue
22	Orange	Violet	Violet	Orange
23	Green	Violet	Violet	Green
24	Brown	Violet	Violet	Brown
25	Slate	Violet	Violet	Slate

the tip and the other as the ring. It is very important to not mix the leads, as there is a difference in the usage in most cases.

PUNCHDOWN BLOCKS

The other end of the twenty-five or fifty pair telco cable is usually terminated into an RJ21 or R66B block. The difference between the two is that an RJ21 is terminated with another Amphenol connector, while an R66B is terminated with a punchdown block, or punch block for short. Some blocks have Amphenol male connectors, while others have female terminations. They are available in either configuration. Punch blocks are also available in multiple configurations with all punches on a horizontal row electrically connected or with a split row, which must have a jumper clip to connect both sides of the split row (see figure 3.3). Newer model R66B blocks are

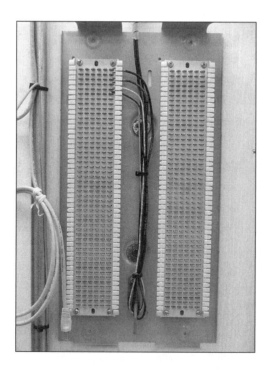

FIGURE 3.3 R66B block

also available, which offer additional features such as Category 5 rating and user installable block interconnect connectors, such as Amphenol connectors or Category 3 and Category 5 jacks.

The 110-punch block is another common interconnect block used in telco and Local Area Network (LAN) wiring. The advantage of the 110 block is a smaller connection terminal or punch area. The terminal can act as an antenna at high data speeds. The smaller terminal reduces opportunities for RF interference. The 110 block uses a vertical stacking connection rather than the horizontal rows used in the R66B. Both the R66B and 110 blocks require special punch tools. Make sure you use the proper tool for the punch block.

WireWrap Block

Many sites also use a WireWrap terminal block for the termination (see figure 3.4). These are used because many wire pairs can be terminated in a relatively small space. There is no standard for the pin-out wiring on a terminal block. Some sites have the number one pin at the front left, others at the back left, and some even use the front right or the back right. The important thing to do is to be consistent and label your block.

Data Communications Interfaces

Several data communications interfaces are used in wireless sites. The most common are the RS-232C and Ethernet. These two standards will be discussed briefly in this chapter. Later chapters will discuss data communications in greater detail.

RS-232C

Equipment manufactured today uses data terminals or computer interfaces which require RS-232C interface cables. RS-232C is a standard that has been around for over thirty years that defines the voltage and pin-outs on data cables. The original RS-232C specification limited cables to lengths less

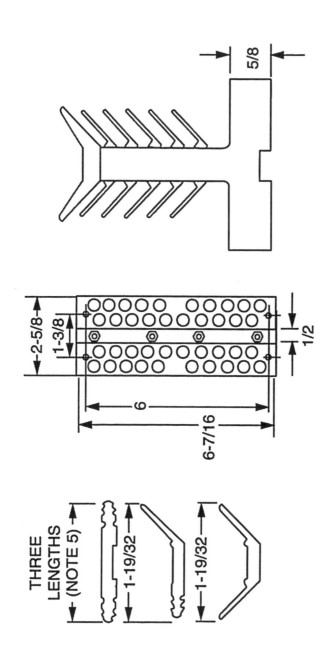

FIGURE 3.4 WireWrap block

CHAPTER 3: EQUIPMENT WIRING

CONCEPT FOR REVIEW:

WIREWRAP FUNDAMENTALS

WireWrap is a method of connecting many wires in a very small space. The most popular place to see this technique in use is in the Main Distribution Frame in a telephone switching office. Here, thousands of wires can be terminated in a very compact manner.

When using WireWrap, there are quite a few precautions that you must follow for the connections to be solid and reliable. Those precautions are listed here:

- Use only the proper stripper for the insulation.
- Use only the proper wire size for the post size.
- Use only the correct size bit in the gun or tool.
- Do **not** nick the wire with the stripper.
- Do not reuse a previously used part of the wire.
- You must have at least five turns on the post.
- Only use square posts.
- Do not leave gaps between wire turns.
- Do not make the wire too taut.
- Put the wire all the way at the bottom of the post.
- Do not reverse tip and ring during the installation.
- You must not strip too much excess insulation, leaving more than 1/8" of bare wire. More than 1/8" of bare wire is called a "shiner". In fact, it is better to have the first full wrap with the insulation attached. This is called Modified WireWrap.
- The installation on the shop side is in the reverse order than on the user side. Be careful, as the count is based on the user side.
- If you do not have enough turns or have a bunched wrap, soldering does make a bad wrap job acceptable.

If all of these rules for WireWrap are followed, a very reliable connection can be made.

CONCEPT FOR REVIEW:

RS-232C PIN-OUT

The specifications for RS232 have three parts:

- Standard connectors
- Defined pin-out connections
- Standard voltage levels

There are two types of standard connectors:

- DB25
- DB9

These can be either male or female in gender.

The standard pin-out for a Data Terminal Equipment (DTE) connection is shown in the table below:

Name	DB25	DB9	Input or Output
TX data	2	3	O
RX data	3	2	I
Ground	7	5	—
Carrier detect	8	1	I
Ready to send	4	7	O
Clear to send	5	8	I
Data set ready	6	6	I
Data term rdy	20	4	O
Ring indicate	22	9	I

(Continued.)

(Continued.)

If the equipment is designed as communication equipment, where the official designation is Data Communications Equipment (DCE), the pin-out is as follows:

Name	DB25	DB9	Input or Output
TX data	3	2	I
RX data	2	3	O
Ground	7	5	—
Carrier detect	8	1	O
Ready to send	4	7	I
Clear to send	5	8	O
Data set ready	6	6	O
Data term rdy	20	4	I
Ring indicate	22	9	O

The voltage levels are +3 to +15 VDC for a high, and −3 to −15 VDC as a low. There is never a level at zero volts, and all of the voltage levels are referenced to the ground lead.

than fifty feet. Longer cable runs can be supported using line drivers or modems. Several methods of handshaking, or control, have developed over the years. Some of these methods are hard-wired; others are software controlled. Software handshaking requires fewer conductors for device communications.

Before calling the small nine-pin connectors DB9 connectors, the series of connectors were called D Sub-Miniature series connectors. The designations and pin count are as follows:

DA15	15 Pins
DB25	25 Pins
DC37	37 Pins
DD50	50 Pins
DE9	9 Pins

After the introduction of the IBM AT computer in the early 1980's, people started calling the DE9 connector a DB9, to match the DB25, which was the connector type used for both the serial and parallel ports on the CPU cabinet. That new designation has stuck as the name of the nine-pin connector.

ETHERNET

Many systems are now connected using Local Area Network (LAN) technology, and these cables must be properly installed. Ethernet is the most common physical LAN interface and it has become very popular as a high speed data communications interface for peripheral equipment. Ethernet has several advantages over RS-232C:

- Higher data speeds
- More compatible devices
- Longer distances possible between devices

Ethernet was first implemented with coaxial cable. This was called 10Base2 Ethernet. Newer implementations of Ethernet operate at ten megabytes per second (10BaseT) or at 100 megabytes per second (100BaseT). While the 10Base2 uses coaxial cable, the 10BaseT and 100BaseT use Category 5 (CAT5) cable (four pair twisted cable). Coaxial Ethernet cables typically use BNC connectors. Twisted pair typically uses RJ-45 connectors. Several standards exist for wiring the RJ-45, including the following:

- EIA/TIA 568A
- EIA/TIA 568B
- USOC

Like RS-232C, some interconnection schemes require different conductor wiring for interconnection. Check your equipment specifications for more information. Both custom fit and standard cables are used at most sites. Custom fit means the installer will have to put on the connectors at the site, while standard means that the installer will be able to use premade cables.

COAXIAL CABLES

Being a radio or wireless site, the one type of cable you would expect to be present is coaxial cable. This cable is used to transport the radio frequency signals as efficiently as possible between the various equipment stages and to the antenna. Radio Frequencies do behave in a known particular fashion, and the different frequencies have different characteristics, so special cables are needed for different purposes. The result is several dozen types of cable, and the necessity of using the proper cable for a given job. We will discuss this more fully in Chapter 6, RF Cabling.

ALARM WIRING

Most radio sites are unattended locations, so there are usually alarm sensors in the site and these must be cabled back to an alarm sending unit (see figure 3.5). The sensor is usually low voltage AC or DC operated, so the wire is usually a pair of small gauge wires (like telephone station wire). The terminations can be screws or special connectors, or you might need special tools for making the connections. There is no standard on these items.

TELEPHONE WIRING

Almost every site will have telephone line for voice communications to the outside world. If you are installing the telephone at a dedicated site, you will also have to install a telephone station. A shared site will have telephone lines, but most of the users do not want anyone other than their own personnel using their line, so the line will be present, but no actual telephone instrument station will be attached. Personnel visiting the site will have to

FIGURE 3.5 A door sensor with wires coming out of it

bring their own phone. The telephone company will normally terminate the line into a RJ11 jack, which is the standard FCC registered interface for the telephone companies to use. They will normally mount this RJ11 block on a wall, but they will run it to anywhere that is specified in the service order, and that does not necessarily have to be a wall. Other types of telephone wiring will be discussed in Chapter 8, Telephone Wiring.

FIBER-OPTIC CABLE

Many sites have Fiber-Optic (FO) cables as one of the interconnecting cable types at the site. Fiber-Optic cables that are run on horizontal racks must be protected in a plastic tube or raceway. Also, Fiber-Optic jumpers come in standard lengths and are usually not cut to fit, so excess cable must be neatly coiled so that it does not interfere with other equipment or persons.

> **CONCEPT FOR REVIEW:**
>
> ## SAFETY ISSUES WITH FIBER-OPTIC CABLES
>
> The wavelengths of the light that the Fiber Optic (FO) communications systems use are outside the visible spectrum of light. As such, a dark fiber, which means that there is no signal present, looks just like an active one at first glance.
>
> The problem that arises is that the energy from a live FO cable can cause blindness if it gets into your eyes. You cannot see the emission from the fiber, but you will discover the effect when you have no vision.
>
> For this reason, always treat **every** Fiber-Optic cable as though it is turned on and just waiting to get your eyes.
>
> Always keep the end covered when it is not terminated into a jack.
>
> The glass fibers inside the cable are also very small, so you should also use safety glasses or goggles whenever you handle FO cable.
>
> Finally, **never** look directly into the end of a Fiber-Optic cable unless you are 1000% sure that the signal is off.

CABLES

It is very important that you always use the proper type of cable for the kind of job you are doing. The engineer for the job will order most of the cables, but occasionally the installer has some input regarding the cable type and the quantity required to make the cabling fit the site. We will look at the actual cables and their characteristics in later chapters, as we discuss the different kinds of jobs.

When the cables are laid in place, many installers use nylon cable ties to dress up the installation. Be very careful to cut off the surplus length of the tail ends of the ties clean at the head. Failure to do so creates a safety hazard. The sharp ends of the plastic will cut anybody who happens to rub against them. There are tools available to properly do this, or you can use flush cutting diagonal pliers if you are careful to cut the end off flush.

FIGURE 3.6   A cable tie with protruding end

CONNECTORS

Just as with the cables, the engineer for the job usually orders the connectors, but the installer must properly attach the connectors to the cables. We will discuss the connectors for various kinds of cables in later chapters.

LABELS

Labels are very important in both the installation and the long-term maintenance of the site. Labels allow the maintenance technicians to remove and replace equipment without retracing each cable. Standards exist for labeling, and some sites have internal labeling standards as well. Here is some of the information often found on a cable label:

- Rack name or number
- Shelf name or number

- Equipment name or number
- "J" or "P" name or number
- FROM and TO information
- Location information
- Part number information

Equipment rooms found on multiple floors in a building will require more elaborate labeling schemes such as TIA/EIA-606. Most installers don't specify this sort of elaborate scheme, but if you encounter it, you can find detailed information on the Internet.

DRAWINGS

The cable drawings are very important for the maintenance and future expansion of a site. An "As Built" drawing should be made for the cabling if the original drawings do not reflect the actual site installation. Some companies require separate drawings for the AC power, DC power, FO, low voltage cables, and coaxial cables.

TESTING THE WIRING

Since cables and connections can have faults, you should perform 100% testing of all installed and premade cables and connectors. Many grades of testing are often specified, with continuity and cross cable resistance tests for conducting cables at basic levels. The highest grades of test include attenuation and actual performance characteristics of the specific cables. Testing can be done either with automated test sets or manually, depending on the level of testing required. Simple continuity tests on copper cables can be done with a DMM or VOM. Attenuation and performance testing may require a Frequency Domain Reflectometer (FDR) or Vector Network Analyzer (VNA).

Summary

Site wiring is crucial to the long-term performance of the site. It is very important that wiring be installed correctly. Failure to do so will result in many maintenance problems later. The proper cable, connectors, and good workmanship are required for every cable installed. The labeling and documentation is also important so that technicians responsible for ongoing maintenance and planning can do their jobs without extensive on-site research.

Many types of cables are used in the wiring of a site. Similar cables can be used for different applications requiring special connectors and installation methods. Some of the interconnection techniques are standard in the industry, others are proprietary to the equipment manufacturer.

Questions for Review

1. To what standard must the AC power wiring be performed?

2. When more than one standard applies, which standard do you use?

3. What is the IR drop if the resistance is 0.04 ohms, and the current is 10 Amperes?

4. Name the five colors and five tracers that make up the telephone cable color code.

5. What color is used for pair six in a twenty-five pair cable?

6. How many turns are required for a WireWrap connection?

7. What is a "shiner" and how long is acceptable?

8. What kind of connector is normally used for a common telephone station?

9. What is the number one safety concern for a Fiber-Optic cable?

10. How do you test a copper cable for continuity?

CHAPTER 4

AC WIRING

OBJECTIVES

After completing this chapter, you will understand the following concepts:

- The National Electrical Code
- Safety precautions to observe when wiring or working on AC circuits
- Licenses and permits required in constructing the AC component of a site
- Methods for determining the amount of electricity that you need to provide
- Methods of providing backup power
- The importance of the load center
- Methods for determining the kind of electrical outlets that are required
- Methods for determining lighting requirements
- The differences between a ground and the neutral line
- Requirements for tower lighting
- Methods of protection against lightning damage
- Methods for testing AC lines and circuits
- Documentation required for the AC component of a site

KEY TERMS

- NEC
- Conduit
- Breaker
- Fuse
- Transfer switch
- Auto start
- UPS
- Lumens
- Neutral
- Surge arrester
- Polarized plug

INTRODUCTION

All radio and wireless sites require power. This chapter will introduce you to some of the common components of the systems that provide power to radio sites. As you can see from the objectives above, there are many facets of the job that relate to Alternating Current (AC) power and wiring.

NATIONAL ELECTRICAL CODE

In the United States, all electrical installations must be wired according to a standard called the National Electrical Code (NEC). This standard was developed by insurance companies, equipment manufacturers, fire chiefs, and engineers to ensure that the electrical wiring in a structure is safe and will not cause fires. Most insurance companies will not accept liability for a claim resulting from a fire if the wiring that caused the fire did not meet this code. Any person who is performing electrical wiring is required to have

CHAPTER 4: AC WIRING

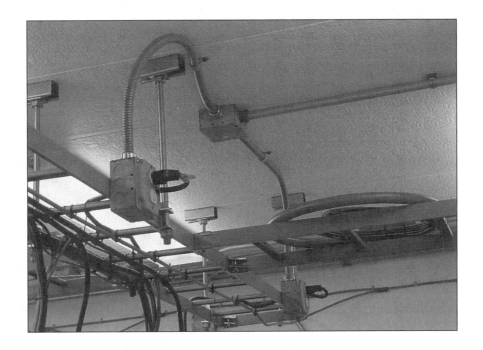

FIGURE 4.1 Wiring in conduit

the NEC at his or her disposal. The NEC can be obtained from the National Fire Protection Association, Inc. (www.nfpa.org), or you can check out a complete series of books on electrical wiring and codes, including the NEC, from Delmar Thomson Learning, the publishers of this book (for more information, visit www.delmar.com).

Some of the key requirements of the NEC are that all wiring must be of a sufficient gauge to prevent overheating, and commercial sites (which includes wireless sites) must have wiring contained in a metal or plastic pipe called conduit (see figure 4.1).

SAFETY

The voltages of the electrical wiring (120 VAC and 240 VAC) are lethal to humans and animals. Obviously, you must take extreme care to ensure that no one is shocked or injured from the electricity at a structure or site. In addition to the standards found in the NEC, common sense should tell anyone to be extremely cautious when working with electrical circuits. If you are not familiar with working with electricity, stay away from exposed circuits. And *always* wear glasses or goggles when working near electricity. No exceptions—*ever*.

LICENSES

Many localities require that anyone connecting electricity to an AC circuit have licensing at a Journeyman Electrician level or above. Some states have added a low voltage section to their codes, so that installing, repairing, or maintaining alarm and other low voltage wiring also requires a state certified electrician.

Many municipalities will not allow the power company to connect the electricity to a site until a building inspector has inspected the site. Consult your local ordinances before installing any AC power at a site.

AC REQUIREMENTS

The AC power to a site must be able to provide enough current to power the equipment at the site. In addition to the radio equipment, this includes the following:

- Lighting
- Tower lights
- Air conditioning
- Heating
- Peripheral equipment
- Outlets around the building

There should also be a reserve factor figured into this amount. The service from the power company must be more than the calculated load, and the

CHAPTER 4: AC Wiring

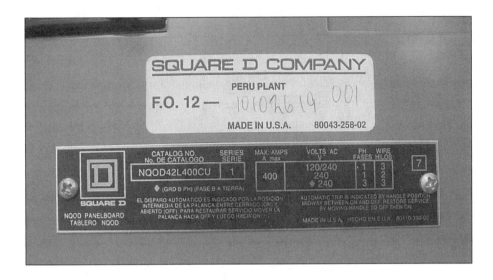

FIGURE 4.2 The plate on a load center showing the max current rating

load center should also have some extra spaces for future expansion. A small site would probably have a service rating of fifty Amperes and a large site would probably have several hundred Amperes of service (see figure 4.2).

BACKUP GENERATORS

Almost all radio sites depend upon AC power to keep the radios running (except for a few very remote sites). No AC service is 100% reliable, so most radio sites have at least one form of backup power. The best way to provide a long-term (two hours to one week) power backup solution is to have a generator as part of the site (see figure 4.3). The generator must be able to supply power to primary equipment, lights, and air conditioning. If a power failure occurs, you may need to do maintenance on the site. You also have to consider lighting and secondary power requirements. Even if power is maintained or restored to the radio equipment, the equipment will still fail if it becomes too hot, so additional power for the air conditioning is required. You have to calculate backup power requirements just like you do the standard AC load requirements.

69

FIGURE 4.3 A generator

An integral part of a generator system is a transfer switch (see figure 4.4). This switch controls whether the power source for the site is the AC line power or the generator. In addition, most transfer switches are programmed with the ability to test the generator's ability to come online. These tests can be performed even though the AC power is still present. The tests can be set to happen automatically at scheduled intervals, or you can run them manually. Only an experienced electrician familiar with generators and transfer switches should install or maintain these items.

Fuel is also required for the generator. Generators require gasoline, diesel, or natural gas as a fuel source. As noted above, AC power failure can last for extended periods, and you'll need to develop a plan for keeping the generator fueled long term. Many sites elect to use a natural gas line with a local storage tank, or pig. This method provides additional levels of redundancy.

CHAPTER 4: AC WIRING

FIGURE 4.4 Transfer switch

LOAD CENTERS

A load center is also called a circuit breaker box. A circuit breaker works like a fuse, but is resetable and can be used as a power cutoff switch. A fuse will fail, or blow, when the current passing through the circuit becomes greater than the rating of the fuse. However, once a fuse fails, it must be replaced; it cannot be reset and used again. A circuit breaker fails, or trips, when the circuit current exceeds the breaker rating. A breaker, however, can be reset and used again without needing to be replaced. A breaker can also function as a manual switch to turn off a circuit for maintenance or testing. Most circuit breakers are tripped by either thermal sensing, or by the heat of the current passing through the breaker circuitry. The way these methods work is complicated, but the only thing you need to worry about as far as

FIGURE 4.5 Load center with cover off

installing them is to be careful if you are selecting thermal circuit breakers that are going to be used in very hot locations.

One side of the load center connection to the circuit breakers *always* has power applied, or is hot, via a common bus bar (see figure 4.5). The other side of the individual breakers are attached to individual circuits and are only hot when the switch is in the **ON** position. You must always exercise extreme care when working in or near a load center.

> **Always wear safety glasses or goggles around a load center when the cover is off.**
>
> **Never perform maintenance on a load center by yourself.**
>
> **Always have someone to assist you in case of an emergency.**

CHAPTER 4: AC WIRING

FIGURE 4.6   UPS

UNINTERRUPTIBLE POWER SUPPLIES

Some equipment cannot ever lose power due to the important nature of the service. Other equipment can be destroyed or damaged if it does not go through a proper power shutdown sequence. For these applications, an Uninterruptible Power Supply (UPS) is necessary (see figure 4.6). UPS systems are battery-powered, and rated according to the maximum current and the length of time that the batteries will provide that current (see figure 4.7). If you also have a generator at the site, the UPS only needs to power the equipment long enough for the generator to come on-line.

Batteries inherently have several safety considerations. Batteries contain caustic chemicals dangerous to humans and equipment. The acids and gels in batteries can produce gasses when charged too much or too rapidly, which can cause the battery to explode. These chemicals can damage your eyes.

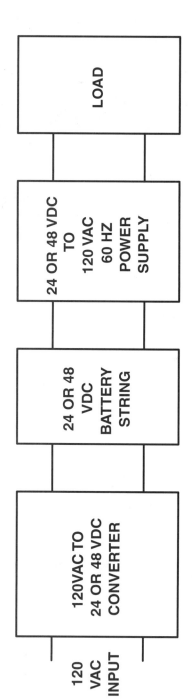

FIGURE 4.7 A diagram of a UPS

CHAPTER 4: AC WIRING

Always wear safety goggles or a face shield when working with batteries. Always store batteries in a well-ventilated room. These chemicals are highly acidic and will quickly damage skin, clothes, or other materials. *Always have a neutralizing agent and fresh-water wash available when working on battery systems.*

A battery is a device that can store power, then release it under controlled or uncontrolled circumstances. Controlled release is the way the batteries are designed to be used. Uncontrolled release of all of the stored energy in the battery may have serious consequences. For instance, accidentally dropping a tool across a battery's terminals can cause it to arc, releasing huge amounts of energy within seconds. Uncontrolled events like this are dangerous to personnel and equipment, and damage the batteries. *Always use caution around unprotected battery terminals.*

POWER OUTLETS

Power outlets are a convenient way to get power to equipment and tools. When outlets are properly installed, the power comes directly from the load center breaker or fuse, through wires enclosed in conduit, to the outlet box. According to the NEC, you must have a three-wire connection between the load center and the outlet (see figure 4.8). These three wires include the following:

- Hot lead — Black
- Neutral lead — White
- Ground — Green or bare wire

Wire connections themselves are normally held captive using the screw or screw lug of the outlet and the breaker or fuse. Because of this, most electricians use solid wire rather than stranded wire. You should make any links, such as connections between AC wiring conductors or junctions between two wires, in an enclosure such as an outlet box or junction box.

The current draw and voltage level (120 VAC or 240 VAC) of a piece of equipment determines the particular outlet type you will need to use. There

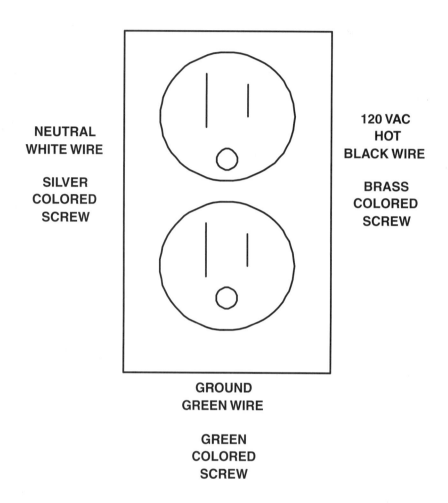

FIGURE 4.8 Close-up of outlet wire connections

is a standard for this detailed in the NEC. Be sure not to exceed the current rating for a particular outlet type.

LIGHTING

At many sites, the lighting for the site is also part of the installation. At other sites, especially shared commercial sites, the lighting is provided by the site manager. As in most residential rooms, the power switch to turn on the lights is located near the doorway entrance to the room. In some cases where sites have multiple exits, three-way switches are used. In these situations, you need to run four wires in the conduit and use three-way electrical Single Pole–Double Throw (SPDT) switches.

Most sites use fluorescent lights for the following reasons:

- Longevity of lamp bulbs
- Higher light output per watt used, greater efficiency
- Even spread of the light output
- Low cost of the fixtures
- Easy to install fixtures

Note: Be aware that there are now two types of fluorescent lamp fixtures and bulbs. If you install the wrong bulb into the wrong fixture, you will destroy the bulb when you first turn on the power. The bulbs look alike, so you cannot easily tell one from the other without reading the printing on the bulb itself. One type of fixture uses 10/20/30/40/50/60/70/80/90/100 Watt bulbs, and only those same type bulbs should be used. The other system uses 8/17/24/34/48 Watt bulbs, commonly called "8's", and only this series of bulbs should be used in these fixtures.

The intensity of lighting is measured in units called lumens. If the lumen level is too low, it is difficult for the maintenance and installation technicians to work at the site. If it is too high, the energy used is just wasted. The engineers should determine the most effective lighting level for their site and situation when they specify the lighting for the building or site. The standard level of light is roughly fifty lumen.

GROUNDS AND NEUTRAL

When the electrician runs wires to the load center, the neutral lead and the ground lead both go to the same connector strip. This means that there should be no potential difference between the ground lead and the neutral lead. Connecting the neutral and the ground together is called bonding and should only be done in the load center. Never bond the neutral and ground together at a remote connection. According to the 2002 NEC, you can now use the outside of a metal conduit as the ground lead. This was not true in some of the earlier issues of the NEC. Many local codes do not allow the conduit to be used as the ground conductor. As a rule, it is still better and required in most situations to have the ground as a separate conductor in the conduit to a plug, light fixture, or equipment termination. Conduit and junction boxes should still be electrically connected to the ground conductor (see figure 4.9).

New nomenclature has been introduced in some areas for AC wiring conductors. Take care not to confuse the nomenclature we have been using in this section so far with the new terms. The new and old names and associated color codes are listed in table 4.1.

TABLE 4.1 Old and new wiring conductor terminology

Old Conductor Name	New Conductor Name	Conductor Color
Hot	Ungrounded Conductor	Black
Secondary Hot	Ungrounded Conductor	Red (used in 220V systems)
Neutral	Grounded Conductor	White
Ground	Grounding Conductor	Green or Bare Wire

TOWER LIGHTS

The Federal Communications Commission (FCC) and Federal Aviation Administration (FAA) require that all towers over 200 feet must be illuminated. The standard for the lights and fixtures is quite rigorous, and only FAA approved lighting equipment can be used. You also must have automatic sensors on the towers to change the lighting mode from daylight to nighttime. Finally, the FAA requires that any time the system malfunctions, it must be reported back to them immediately, so the tower light equipment must include sensors and senders to provide this function.

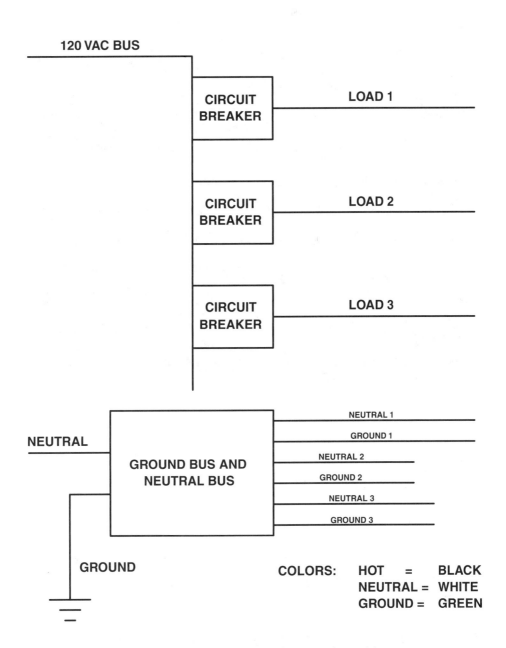

FIGURE 4.9 The ground, neutral, and hot leads from the load center to the outlet

FIGURE 4.10 Tower light controller box

Some tower lights operate on 240 VAC, while other use 110 VAC. Also, the leads are usually exposed once the tower light control box is opened, so be extremely careful when working on or installing this equipment (see figure 4.10).

All tower lighting systems must have a photocell to determine when the night lighting must be on (see figure 4.11).

LIGHTNING PROTECTION

Lightning is always a menace to power lines. There are two main types of lightning protection methods. The first type consists of gas discharge modules that conduct, or fire, when the voltage reaches a few hundred volts

FIGURE 4.11 A photocell

above normal. This is applicable for lights and some motors, but much of the equipment found in a wireless communications site would be destroyed by the time the voltage reached a high-enough level to cause the gas tube to fired. In order to correct this, the second type of protector was developed, which uses solid-state devices that become active if the voltage rises even a few volts for more than a few milliseconds (see figure 4.12). The better, and more expensive, units operate faster than the less expensive units. And as I mentioned earlier in the book, a good grounding system is fundamental and imperative for lightning protection to function properly.

TESTING

AC wiring can easily be tested by a number of means. The important thing to remember is that mistakes can be very dangerous, even fatal. Always wear eye protection whenever you are handling AC wiring in any way.

FIGURE 4.12 Lightning protector

Two instruments that are standard in any test kit are the Digital Multimeter (DMM) and the outlet tester. The DMM is a good instrument for checking voltage on hot leads, continuity on ground or neutral leads, and potential difference on unknown leads. An outlet tester has LED indicators to check that the wiring is correct.

Occasionally, other factors must also be tested on AC systems including decreased voltage, or dropouts, voltage increases, or surges, and line frequency. All of these factors may occur randomly and intermittently and consequently require that the line be monitored over extended periods of time. Dropouts may occur when large electric motors or other large electric loads initially turn on. Large motors require a great deal of energy to begin rotating, and capacitors require energy to charge. Once these surge requirements are satisfied, the line voltage returns to normal. You may have experienced this voltage dropout when your home refrigerator or air conditioner turns on. The lights dim momentarily and then return to normal. These events are typically rapid, lasting a second or less, and may cause intermittent equipment failure. Long-term, over several minutes, low voltage will damage motors and other equipment. Surges may occur when large motors or electrical loads are turned off. This is the reverse of the dropout situation.

CHAPTER 4: AC WIRING

Dropouts or surges occur over several cycles, or wavelengths, of alternating current. Spikes can occur from extremely quick acting events, milliseconds or less, such as a switch from one power supply to another. Spikes take place during the alternating current cycle. Rapid switching devices such as relays and events like switching power supplies can cause electrical noise, which means spikes appear in the normally clean alternating current sine wave. Severe noise can damage equipment.

Improper line frequency can also cause equipment failure due to load/supply mismatches, or because the equipment is designed to detect 60 Hz. When the frequency is too high, the average power increases and causes a situation similar to voltage surge. When the frequency is too low, the average power decreases and may cause equipment failure. Power utility companies go to great lengths to control standard line voltage and frequency. However, line frequency problems are a common point of failure on backup power systems such as generators and UPS systems. Test the voltage and frequency of generators regularly.

DOCUMENTATION

The most important documentation you need for a load center is a list on the breaker box cover that shows what load each breaker controls (see figure 4.13). Also, many sites require that the outlets, lights, and equipment to also have labels specifying which breaker is used for that circuit. This greatly simplifies maintenance and troubleshooting.

SUMMARY

Functional, safe AC wiring is crucial to the operation of most wireless sites, and there are a number of things that you'll need to remember any time you have to install, repair, or maintain the electrical system. The National Electrical Code is the standard that all electrical wiring is required to meet. In some cases, there are even more specifications that must be met. Safety must always be at the forefront of every activity around electrical wiring. Many localities require permits, licenses, and inspections for the AC wiring

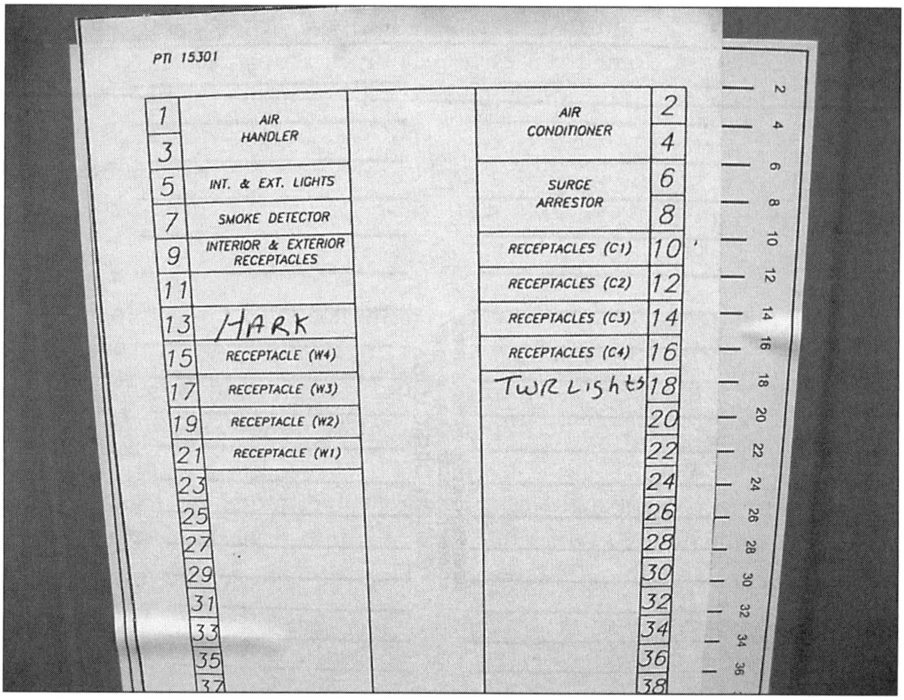

FIGURE 4.13 Load center label

to be completed. The sum total load of the building or site determines the service level and components needed to power the site. Both generators and UPS systems are part of the backup power plan for most sites. The current draw of a piece of equipment determines the type of outlet used on the equipment. The lighting requirements are determined by the engineers in the planning stages of a site. The ground and neutral leads in a load center box are at the same potential. A tower over 200 feet in height is required to have tower lights, which must operate in a day mode and night mode, depending on the ambient light that is outside. Lightning is always a menace at a radio site, so protection must be utilized to protect the equipment in a site. The AC wiring can be tested in multiple ways at a site before the equipment is powered up. The most important documentation to do at a site in regards to the AC wiring is to clearly identify what each breaker does and what circuits are on each breaker.

CHAPTER 4: AC WIRING

QUESTIONS FOR REVIEW

1. What does NEC stand for?

2. What should you wear when working with any AC wiring?

3. Name at least four items that use 120 VAC power at a site.

4. Name the three basic colors used for 120 VAC wiring and what each lead represents.

5. What does a transfer switch do?

6. What are the three common colors used on the insulation of 120 VAC wiring?

7. What does "8's" mean when you are talking about fluorescent lamps?

8. What tower height requires tower lighting?

9. What are the two operating modes for tower lights?

10. What is the most important documentation with respect to the AC wiring at a site?

11. What does an outlet tester do?

CHAPTER 5

24 VDC AND 48 VDC WIRING

OBJECTIVES

After completing this chapter, you will understand the following concepts:

- Safety precautions you need to keep in mind
- The various components of a 24 VDC or 48 VDC power system
- How batteries are maintained
- How the power is distributed at the site
- What the capacity requirements are for various kinds of wiring
- Methods for testing 24 VDC or 48 VDC wiring
- Required documentation

KEY TERMS

- Battery
- Ground
- Float
- Grasshopper fuse
- NO-OX

Introduction

Many wireless sites use 24 volts Direct Current (VDC), or 48 VDC as the primary power source for the site. This chapter details what is involved with a 24 or 48 VDC power system. In addition, we will cover care of the batteries, and provide an overview of all of the components of the power system, which includes the batteries, charger, power distribution panel, fuse panels, cabling, and alarm system (see figure 5.1).

Note: The polarity of a 48 VDC power system is usually set such that the negative lead is the power lead, and the positive lead is tied to ground. This may or may not be true in a 24 VDC system. Be sure to check the manufacturer's requirements before you start any wiring.

Safety Precautions

Even though 24 VDC or 48 VDC is not enough to cause a shock if you touch the power source, there are precautions you need to take when working with these systems. The danger and safety precautions you need to take are about preventing short circuits that are caused by a metal conductor crossing the battery supply. Because the voltage is not very high, there are many exposed places in the circuit. But the currents can be very high, and a short circuit will vaporize the metallic conductor and shoot extremely high temperature sparks from the short. For this reason, *always* wear glasses or safety goggles when working around battery supplies and systems.

Battery String

The battery string in a wireless site is usually one of two types (see figure 5.2). The first type is the lead-acid 2.0-volt cells that are connected in series to produce 24 VDC or 48 VDC. These are expensive, hard to maintain, and produce hydrogen gas when charging and heavy. They are used where extremely high currents, in the order of hundreds of Amperes, are required. For most sites, the current draw is less than 100 Amperes and the batteries

CHAPTER 5: 24 VDC AND 48 VDC WIRING

FIGURE 5.5 A power cable on an overhead rack

POWER DISTRIBUTION PANELS

Once the battery voltage is brought to a rack or cabinet, it is usually terminated into an indicator fuse power distribution panel (see figure 5.6). From here, there are smaller gauge leads that bring the power to the individual shelves in the bay.

FUSES AND CIRCUIT BREAKERS

Most 24 VDC or 48 VDC circuits have at least two circuit breakers or fuses per circuit. A primary fuse or breaker is attached to the feeder cable right at the main power board. This fuse or breaker is usually fifteen Amperes to fifty

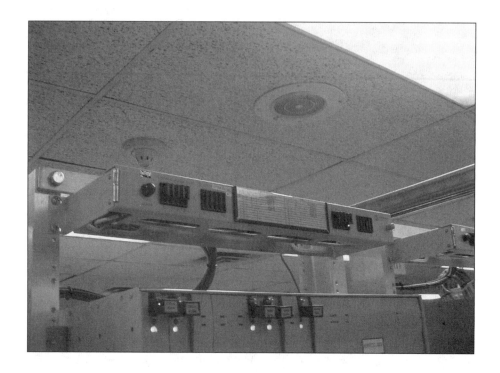

FIGURE 5.6 A power distribution panel

Amperes per circuit. There is always an indicator fuse here too, which will blow any time the main fuse blows. When this happens, the mechanical design of the indicator fuse allows the voltage to make to an alarm circuit path, generating a fuse alarm at the site and alerting the proper personnel that a fuse has blown. Many indicating fuses look like grasshoppers, so they are called grasshopper fuses. The indicator fuse will have a very low current capacity, so it will always be blown when the main fuse is blown. The indicator fuse is wired in parallel with the main fuse or breaker to alert the maintenance technicians of a blown fuse or tripped breaker (see figure 5.7 and figure 5.8).

The second fuse per circuit is located in the bay or in the cabinet at the top of the bay. This fuse is usually another indicator fuse, but it is rated for twice the actual current of the circuit it is trying to protect.

CHAPTER 5: 24 VDC AND 48 VDC WIRING

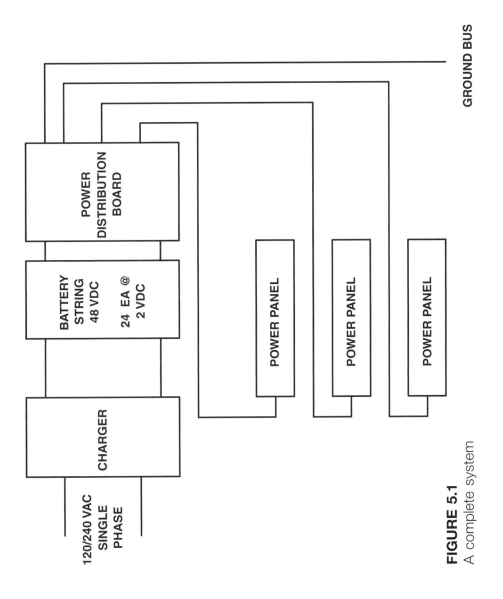

FIGURE 5.1
A complete system

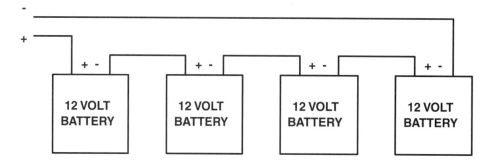

FIGURE 5.2   Battery String

most popular for these sites are gel electrolyte, commonly called gel-cells. This type of battery is a giant version of the backup batteries found in your home burglar alarm or in the UPS you might have attached to your computer. The design engineer typically orders a gel-cell string that totals 24 or 48 volts at the required current capacity. These gel-cell batteries require no maintenance, other than normal replacement every five years. Many engineers are unaware of the time limit on these batteries, and do not realize that they have stopped working until the site suffers a power outage.

CHARGER

The charger takes the 120 VAC or 240 VAC from the power company and converts it to DC to charge the battery string (see figure 5.3). Once the voltage of the batteries is up to the fully charged level (26.2 or 52.4 volts), the charger switches from a full current mode to a trickle charge mode and the main current draw for the site's equipment is from the batteries. At this point, the batteries are in a float mode. Many sites have redundant chargers, so that if one breaks or malfunctions in any way, there is a backup available on-site.

As noted earlier, lead-acid batteries generate hydrogen gas when charging. Gel electrolyte batteries produce gas also, though not to the degree that lead-acid batteries do. Overcharging of either battery type can be dangerous. Excessive hydrogen gas produced by lead-acid batteries can be explosive. Gel-cell batteries will deform from overcharging and may burst their

FIGURE 5.3 Battery charger

containers. No battery should be charged when it is frozen or when the temperature exceeds the battery manufacturer's specifications.

As more equipment is added to a site, many engineers forget to look at the charger and battery string to be sure that the string can accommodate the added load. Be sure to review this facet whenever new equipment is added to a site.

Connectors

In a 24 VDC or 48 VDC system the current can be very high. Following Ohm's Law, the voltage drop described in the formula "current times resistance" will be substantial if the resistance is even slightly elevated. Because of this, the connectors must be extremely low in resistance. Not only must the connection resistance be low from the start, but the resistance must stay

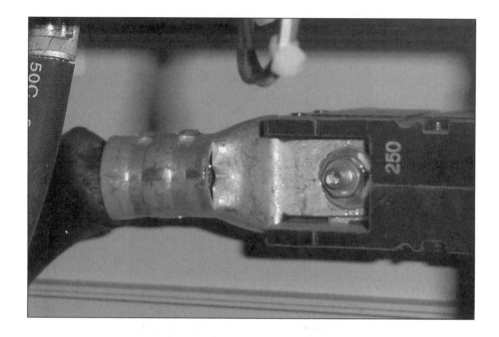

FIGURE 5.4 A close-up of a power connector

low as the connection ages. In order to ensure this, many companies require that a material called NO-OX, which prevents oxidation, be applied to the wires and connectors prior to making the connections and connectors. To further lower resistance at the point of contact, the connectors usually have more than one point of contact and are silver plated (see figure 5.4).

CABLES

Not only are the connectors designed for low resistance, but the cables are also selected on a per job basis so that the total voltage drop of the cables and connectors does not exceed 1.0 volt from the battery to the equipment.

The cables are secured in place on overhead racks (see figure 5.5), on wall brackets, or below the floors so that there is a semblance of order in the running of the power cable.

CHAPTER 5: 24 VDC AND 48 VDC WIRING

FIGURE 5.5 A power cable on an overhead rack

POWER DISTRIBUTION PANELS

Once the battery voltage is brought to a rack or cabinet, it is usually terminated into an indicator fuse power distribution panel (see figure 5.6). From here, there are smaller gauge leads that bring the power to the individual shelves in the bay.

FUSES AND CIRCUIT BREAKERS

Most 24 VDC or 48 VDC circuits have at least two circuit breakers or fuses per circuit. A primary fuse or breaker is attached to the feeder cable right at the main power board. This fuse or breaker is usually fifteen Amperes to fifty

FIGURE 5.6 A power distribution panel

Amperes per circuit. There is always an indicator fuse here too, which will blow any time the main fuse blows. When this happens, the mechanical design of the indicator fuse allows the voltage to make to an alarm circuit path, generating a fuse alarm at the site and alerting the proper personnel that a fuse has blown. Many indicating fuses look like grasshoppers, so they are called grasshopper fuses. The indicator fuse will have a very low current capacity, so it will always be blown when the main fuse is blown. The indicator fuse is wired in parallel with the main fuse or breaker to alert the maintenance technicians of a blown fuse or tripped breaker (see figure 5.7 and figure 5.8).

The second fuse per circuit is located in the bay or in the cabinet at the top of the bay. This fuse is usually another indicator fuse, but it is rated for twice the actual current of the circuit it is trying to protect.

CHAPTER 5: 24 VDC AND 48 VDC WIRING

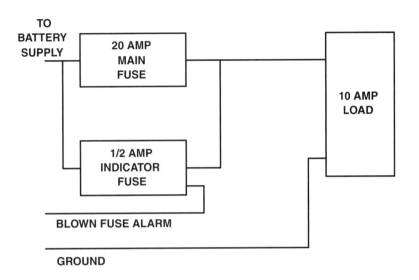

FIGURE 5.7 Main fuse and indicator fuse

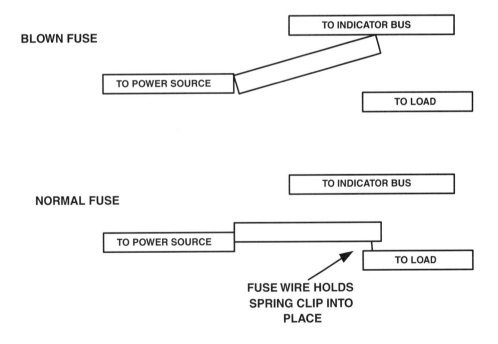

FIGURE 5.8 How an indicator fuse works

Alarms

24 VDC or 48 VDC power systems have many alarms for monitoring the status of the system. These alarms can indicate a number of problems:

- Defective charger
- Over-voltage
- Low voltage
- Blown fuse or tripped breaker in power board
- Blown fuse or breaker in bay

Testing

You will do different tests on a 24 VDC or 48 VDC system during installation than you will during the maintenance phase. During installation, you will need to test polarity and continuity. Polarity testing is just using a voltmeter to insure that the leads are not reversed, and that positive is positive and negative is negative. It is very important that the polarity is correct, as reversed polarity can seriously damage equipment. Continuity testing is just checking that the correct leads are in the correct position, and the resistance of all of the connections is very low.

Once the system is up and running, you'll need to test the following:

- Voltage of the battery string
- Voltage of the string under load with the charger off
- Voltage drop of the cable as measured at the load
- Battery condition and age
- Tightness of the bolts on each junction
- Current draw of the load

CHAPTER 5: 24 VDC AND 48 VDC WIRING

> **CONCEPT FOR REVIEW:**
>
> **VOLTAGE DROP OF A TWENTY AMP CIRCUIT WITH 0.06 OHMS RESISTANCE**
>
> If you have a circuit that is drawing 20 Amperes, and the total resistance of the conductors and connectors between the power source and the load is 0.06 ohms, the voltage drop will be 1.2 volts. This exceeds the 1.0 volt maximum allowed in most systems for the IR drop of the conductors and connectors.
>
> $$E = I \times R$$
>
> $$1.2 = 20.0 \times 0.06$$

For the most part, you do not need anything more complex than a DMM to test a 24 VDC or 48 VDC power plant. Any voltage drop that exceeds 1.0 volt is considered too high, and you'll have to find and correct the cause. Your system must not have a voltage drop greater than 1.0 VDC as the power moves through the fuses or circuit breakers and cabling.

Documentation

The most important part of the documentation for a power board or fuse panel is identifying what each fuse or breaker goes to. In addition, the size of each main fuse and each pilot fuse should appear on the equipment, stenciled or labeled in some way, in easy view for the maintenance technicians. Send a hard or soft copy of all this information to the system engineer for the site records. This documentation is required during the installation phase of a project. The battery condition is the most important part of the documentation in the maintenance phase of a system. This maintenance documentation should detail the following items:

- The age of the battery, with a note attached that reminds personnel to replace the battery sometime after four and before five years of use

- The voltage of each cell
- A physical description of each connector, indicating whether corrosion is forming
- The temperature of each battery
- Any signs of leaking acid or gel material

Summary

24 VDC or 48 VDC is used as the primary power source for the equipment at many sites. There are several issues you need to be aware of when working with this kind of system. The 24 VDC or 48 VDC system consists of the batteries, chargers, power distribution boards, cables, fuse panels, and the alarm system. The batteries are normally the gel-cell type and must be rated to provide a capacity to match the site. They also have an expiration date and must be replaced every five years. The voltage drop of the power distribution must be kept below 1.0 volts in total. To do this, all of the connectors and cables must have very low resistance. Even a small amount of resistance will show up as a high voltage drop. There is almost always an alarm that should go off whenever a fuse is blown or a breaker is tripped. There are other alarms that report any malfunction of the battery system. Testing a 24 VDC or 48 VDC system is relatively simple. Test for correct polarity in the installation phase, and check battery float voltage and drop in the maintenance phase of system operation. The documentation that is most important is the designation and size of each fuse, and it needs to be clearly labeled on each panel near the fuse or breaker. You should also keep a log of battery condition.

CHAPTER 5: 24 VDC AND 48 VDC WIRING

QUESTIONS FOR REVIEW

1. What are the components of a 24 VDC or 48 VDC power system?

2. What types of batteries are used in a 24 VDC or 48 VDC power system?

3. How often should gel-cell batteries be replaced?

4. Where are the power distribution panels usually located?

5. Describe how an indicator fuse works.

6. How is a fuse alarm generated?

7. What should the maximum voltage drop be on a given circuit?

8. What instrument is used for testing voltage drops?

9. What documentation is required for a 24 VDC or 48 VDC power system?

10. On a 48 VDC power system, which lead, + or -, is connected to the ground potential lead?

CHAPTER 6

RF CABLING

OBJECTIVES

After completing this chapter, you will understand the following concepts:

- The elements involved in radio frequency (RF) cabling at a site
- The safety precautions that must be observed when working with RF systems
- The main parameters for choosing components for RF cabling at a site
- The components available for RF cabling
- The lightning protection that is available for RF cabling
- The testing that is required to confirm that the RF cables are properly constructed

KEY TERMS

- Coaxial cable
- Radiation
- Egress

- Ingress
- Shielding
- Leakage
- Loss
- Attenuation
- Shunt
- DC blocking
- MPE
- Decibels

INTRODUCTION

This chapter deals with Radio Frequency (RF) cabling at a site. There are many configurations to select from, and many parameters that dictate what components you'll need. The installation is very important; there is no room for error in the installation and construction of the cables and connectors. In order for a site to look good, many of the cables are made on-site, which puts a greater burden on the installer to do the job properly and to thoroughly test each component. This chapter will serve as both a guide and reference for these components.

SAFETY PRECAUTIONS

RF energy has long been known to cause physical danger to humans and animals. You would not put a dog, cat, or yourself into a microwave oven, but that is exactly what you are doing to yourself if you are in close proximity to enough RF energy at a wireless site. A burn from RF energy starts deep within your body, and not until it has "cooked" you to the outside skin, where your pain sensors are located, do you realize that you

CHAPTER 6: RF CABLING

have been exposed to a high level of RF energy. Your eyes are extremely sensitive to even low radiation levels, so always be aware of what is hot and what is not. You will not notice this damage immediately, but the long-term effects are irreparable. Do not ignore this warning.

The term MPE means Maximum Permitted Exposure. You can get a 100+ page document that defines MPE and explains all the ways it should be implemented from the FCC web site, but for installation purposes, what you need to remember is that the aggregate energy from all of the transmitters at a site combined should not exceed the MPE level. A loose shield, unterminated cable, or improperly made cable will allow RF to be radiated into the area around the cable. There are sensors that you can wear with alarms that sound when the RF energy is high enough to cause physical harm.

CABLE LOSS

All RF cables will allow some loss of the signal as it travels through the cable. This is due to resistance losses and capacitance between the conductors and the inductance of the cable. As the frequency increases, so does the loss for a given cable. You'll need to use a lower loss cable to keep the signal level where it is designed to be as the length of the cable you are using increases. All cables have a loss rating expressed in decibels per 100 feet, and the values are given in a chart form for set frequencies. Table 6.1, in the next section, is a sample of this.

When the frequency is above 3.0 GHz, waveguide is used instead of coaxial transmission line because the loss factor is too great at those frequencies with coaxial cable.

CABLES

There are many types of cables and many manufacturers for each type of cable. The physical characteristics of the various types help the system design engineer determine what cables to order for the major cable runs, but the

> **CONCEPT FOR REVIEW:**
>
> ## COAXIAL CABLE LOSS
>
> As a signal travels through a cable, it loses some its strength. This is called attenuation. The factors affecting signal attenuation include the following:
>
> - Frequency
> - Length of the cable
> - Size of the cable
> - Material that the cable is made of
> - Shielding factor of the cable
>
> Many cables have a rating of the loss per foot, or per one hundred feet, at a given frequency. Regardless of the kind of cable you chose for a particular run, the longer the cable, the more loss you will have.
>
> The losses that occur are due to resistance of the conductors, capacitance between the conductors, the material used for the insulation between the conductors, and the leakage of the signal from the cable.
>
> If the loss is too great, you have the option of trying several things:
>
> 1. Increase the signal level going into the cable.
> 2. Use a larger cable with a lower loss rating.
> 3. Change the system to require less cable.
>
> Sometimes, it takes a combination of all three options to get a signal from Point A to Point B while maintaining the correct signal level.

installer usually provides and installs the short runs and jumpers. To make sure that the minor cables the installer chooses are compatible with the ones chosen by the engineer, the installer must have the same information that the engineer has access to. This is provided in table 6.1.

The impedance of a cable is very important. In a radio site, most of the RF cables have an impedance of 50 ohms. There are a few exceptions, and these include the cables going to GPS antennas and to satellite head-end converters, which are at an impedance of 75 ohms. With the interference found at most sites today, almost all of the manufacturers and system engineers are specifying RG-6 for these applications, because it has 100% shielding.

You will want smaller cables for the following uses:

- Interstage signals
- Signals traveling from transmitters to combiners
- Signals traveling from receivers to multicouplers
- Clock synchronization signals
- Low frequency oscillators

You will need larger cables for these purposes:

- Signals traveling from combiner output to lightning protector
- Signals traveling from lightning protector to antenna
- Antenna jumpers

Some systems use one antenna per system, while others use sectorized antennas, which means that there will be multiple coaxial lines running up the tower.

Even though coaxial cable is constructed with an inner conductor surrounded by an outer shield, there can still be leakage of the signal into or out of the cable. Signals that escape from the cable are called egress, while foreign signals that get into a cable are called ingress. This is definitely an undesirable characteristic for any coaxial cable, so be sure to choose cables with 95% or better shielding to prevent egress and ingress problems.

There are many manufacturers and types of coaxial cables in use today. Table 6.1 lists some of the characteristics of the smaller coaxial cables.

TABLE 6.1 50 ohm Coaxial Cable Specifications

Type	VF in %	\multicolumn{7}{c}{Loss in dB per hundred feet (If specified) at (MHz)}

Type	VF in %	1	10	100	400	900	1000	2000
RG-8	66-84	0.1	0.5	1.6	3.2	5.7	6	
RG-8X	80-84	0.2	0.9	2.8	8	12.8	14.3	
RG-9	66	0.2	0.6	2.1			8.2	
RG-58	66	0.3	0.4	4.5	8.5	13.0	14.3	
RG-58A	66-78	0.4	1.5	5.4	12.4	21	22.8	
RG-58C	66-83	0.3	1.0	3.2			10.5	
RG-142B	69.5	0.3	1.1	3.9	8.2	12.5	13.5	
RG-174	66	1.9	3.3	8.4	17.5	28.2	34.0	
RG-213	66	0.2	0.6	2.0	4.1	7.6	8.2	
RG-214	66	0.2	0.6	1.9	4.8	7.6	8.0	
RG-400	69.5	0.4	1.1	4.5	10.5		13.2	
LMR400	76-85	0.1	0.4	1.3	2.5	3.9	4.1	
LMR500	86	0.1	0.3	0.9	2.1	3.1	3.3	4.8
LMR600	86	0.1	0.2	0.8	2.0	2.9	3.0	3.9

Sometimes you'll be using the cable to provide power to active components such as down-converters or preamplifiers. In these cases, you may need to add DC blocking adapters to prevent the DC signal from reaching components such as combiners or shunt-type lightning protectors, which cannot tolerate DC or which will short out the DC source.

CONNECTORS

Just as there are many types of cables, there are also many types of connectors. There are also many manufacturers of connectors. The connector must match the cable. The following photographs illustrate many of the types of connectors found in the field today (see figures 6.1-6.18).

CHAPTER 6: RF CABLING

FIGURE 6.1   7/16 DIN connector — male

FIGURE 6.2   7/16 DIN connector — female

Sometimes you need a male connector, sometimes a female, and sometimes you need to convert the sex or series from one type to another. When installing or maintaining a site, you need to have a complete set of interseries connectors. These can be discrete connectors or a special interseries connector kit.

You will have to solder some connectors to the cables, while others are crimped into place. The connection method is sometimes specified by the client or manufacturer, and sometimes it is left to the discretion of the

FIGURE 6.3 BNC series connector — male

FIGURE 6.4 BNC series connector — female

CHAPTER 6: RF CABLING

FIGURE 6.5 UHF connector — male

FIGURE 6.6 UHF connector — female

FIGURE 6.7 Type N connector — male

FIGURE 6.8 Type N connector — female

CHAPTER 6: RF CABLING

FIGURE 6.9 RCA connector — male

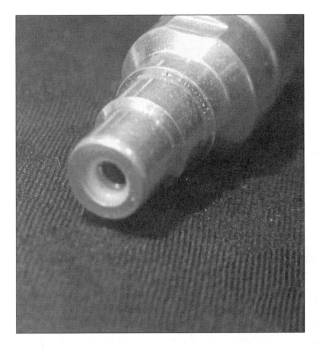

FIGURE 6.10 RCA connector — female

FIGURE 6.11 TNC connector — male

FIGURE 6.12 TNC connector — female

CHAPTER 6: RF CABLING

FIGURE 6.13 SMA connector — male

FIGURE 6.14 SMA connector — female

FIGURE 6.15
Mini-UHF connector — male

FIGURE 6.16
Mini-UHF connector — female

CHAPTER 6: RF CABLING

FIGURE 6.17 "F" series connector — male

FIGURE 6.18 "F" series connector — female

installer. Whether you are crimping or soldering, you need to have the right tools and jaws available. Be sure you know what is needed before you go to the site.

The outside plating on the connector can be silver, gold, nickel, copper, or brass. The lower the resistance, the better the connection. You never see gold at a site, as it will always become a target for thieves, so silver is the material that is normally used for good connectors. Brass is sometimes used, but it has more loss, and copper corrodes, so it is seldom used.

The varieties of cable, connector, plating, insulator, and manufacturers will mean that there are thousands of combinations possible, and thousands of part numbers you'll need to sort through whenever you need to order connectors.

JUMPERS

Jumpers are used to interconnect system components where larger cables cannot bend or fit into the required space. Jumpers normally do have more loss than the larger cables, but the runs are usually short, so the loss is insignificant. In fact, any connector has greater loss than a few feet of any type of cable. Jumpers can either be premade or custom to fit at a site.

LIGHTNING PROTECTORS

Lightning can wreak havoc with a radio communications system. Even a lightning protector cannot protect a system from the massive amounts of power and damage that come from a direct hit. Fortunately, most lightning hits are induced voltage spikes, and lightning protectors do their job of keeping the equipment from being damaged. The high voltage and current from a lightning strike anywhere in the area, even if it just hits the ground, will cause all of the leads in close proximity to have a high voltage spike. This is called induction. If these leads are tied to a lightning protector, the voltage spikes can be managed. The lightning protector is tied to the building ground with a good cable and proper connectors.

There are three types of protectors:

- Gas tube
- Series fuse
- Shorted quarter wave stub

The site engineer should determine which type of protector to use based on the requirements of the site.

Gas tube protectors work by shorting out the transmission line when the voltage on the line exceeds a few hundred volts. Gas tube protectors may not work quickly enough for some delicate electronics.

Fuse-type protectors work by having the fuse blow when the current exceeds a few extra milliamps or Amps, depending upon the power rating of the protector.

Shorted quarter wave stub antennas are frequency dependent, and work by having a DC short between the outer conductor and the inner conductor. At the resonant frequency of operation, the short looks like an open, and does not impede or shunt the operational frequency. There is more detailed information about shorted stub antennas in Chapter 10, Lightning Protection.

TESTING

RF cables can be tested in three different ways:

- Continuity testing with a Digital Multimeter (DMM) or Volt-Ohm-Meter (VOM)
- Testing with a wattmeter
- Testing with a Time Domain Reflectometer (TDR)
- Testing with a Frequency Domain Reflectometer (FDR)

Of the four methods cited above, the FDR method is the most accurate and dependable way to confirm that a coaxial cable is made correctly. It can also

> **CONCEPT FOR REVIEW:**
>
> ## TIME DOMAIN REFLECTOMETRY (TDR)
>
> Time Domain Reflectometry is a method used to test the length of a cable and verify that there are no shorts or opens on that cable. The way that this works is for a very short DC pulse to be injected into one end of the cable being tested. The length of time that the pulse takes to propagate from the origin point to the end of the cable and then for the reflection to come back to the origin point is measured and displayed on a screen or graph print-out.
>
> Electrical and electronic signals travel at the speed of light (186,000 miles per second) in free space, but inside a cable this speed is slowed by a certain amount. This reduction in speed is called the velocity factor. You can usually find the velocity factor for the particular cables you are using by looking in tables provided by the manufacturer.
>
> Given the time that it should take for a pulse to travel up and back on the cable, based on the velocity factor of the cable, the TDR will show how long a cable is and if there are any opens or shorts on the cable.
>
> The shortcomings of a TDR sweep are that minor disturbances of the impedance of the line and frequency sensitive components are not shown properly on the display output.
>
> These shortcomings are corrected in a technology called Frequency Domain Reflectometry (FDR).

be used to verify that the adapters and connectors are made correctly, and is the only method that truly checks that the antenna is working properly.

Summary

If RF is the lifeblood of wireless systems, RF cabling is the circulatory system. There are a number of things to remember when installing and maintaining this vital piece of the wireless site. RF energy can be harmful to

CHAPTER 6: RF CABLING

CONCEPT FOR REVIEW:

FREQUENCY DOMAIN REFLECTOMETRY (FDR)

When a coaxial transmission line is used to transport radio signals from the transmitter to the antenna, from the antenna to the receiver, or from one point to another point, the best way to ensure that the transmission line is working to design specifications is to perform a Frequency Domain Reflectometry sweep of the line. In addition to testing cables, a FDR sweep can be performed on most RF components, such as antennas, filters, lightning protectors, etc., to insure that they are working properly.

The principle that FDR uses is the fact that all components within a transmission line and antenna system are designed to work at an impedance of 50 ohms. The FDR sweep shows the amount of signal

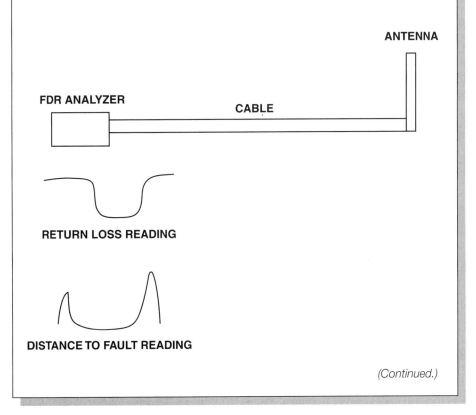

(Continued.)

> *(Continued.)*
>
> amplitude reflected at many discrete frequencies as the device or line is tested on and near the design frequency of operation. It does this by reading the amount of reflected signal at the signal source end, and making an amplitude and time measurement for the reflected signal. This will let the operator of the FDR analyzer see the amount of reflection with respect to frequency, and the distance from the origin point for these reflections.
>
> A perfect match will have a very low return echo. A poor match will have a very high return echo. The FDR analyzer will show the exact level of this return echo.

people, so always observe safety precautions. Cables have loss, so be sure to take the frequency and line length you require into consideration when you are deciding what cable type to use. It is also important that the connectors are the proper type for the job, and that they match the cable. The material that the connector is made of and the attachment mode also helps determine the exact cable connector you should use. Use jumpers in tight, confined spaces where the bigger cables are hard to work with. Lightning protectors prevent lightning induced voltage spikes from destroying the equipment. Cables need to be tested, and the best way is a method called Frequency Domain Reflectometry.

CHAPTER 6: RF CABLING

QUESTIONS FOR REVIEW

1. What does MPE mean?
2. What is egress?
3. What is ingress?
4. What is the relationship between frequency and loss for a given cable?
5. How is the loss of a cable usually expressed?
6. Name at least six types of RF connectors.
7. Why are jumpers used?
8. Name three types of lightning protectors.
9. How can cables be tested to confirm that they are made properly?
10. What is the difference between a male and a female connector?
11. What is the best method to check an antenna installation?

CHAPTER 7

ANTENNA INSTALLATION

OBJECTIVES

After completing this chapter, you will understand the following concepts:

- The elements of antenna installation
- Safety precautions that must be undertaken with an antenna installation
- The different types of antennas
- The different types of mountings
- The different ways to install antennas
- The connectors that are used
- Methods of protecting the cables during the installation and afterwards
- Methods of testing the equipment that comprises the antenna system

KEY TERMS

- Galvanized
- Harness

- Grounded
- Dipole
- Drip hole
- Polarization
- V-Bolt
- Frequency Domain Reflectometry
- Polarization

INTRODUCTION

The most important part of any radio or wireless system is the antenna. No other piece of equipment has a greater effect on the range of a radio system than the antenna. If the antenna does not perform as designed, the signal range will be less than expected and required. One of the criteria any antenna design must meet is the need to eliminate interference, both to the internal system and to external users of other systems. If all of the antenna parameters are not correct, there can be problems for both.

The antenna is normally installed on a radio tower or a building rooftop (see figure 7.1). The antenna is exposed to the elements every day of the year, so it has to hold up in high and low temperature extremes, as well as survive ice, snow, wind, and rain. These harsh environmental elements must be the determining factor in the quality and workmanship of the installation.

DISCLAIMER ON INSTALLATION AND CLIMBING

Only a qualified antenna installer should ever be used or hired to do the installation of an antenna. There is special insurance for this occupation, and there are also special riders and policies for the commercial liability insurance required by all companies in the radio business.

CHAPTER 7: ANTENNA INSTALLATION

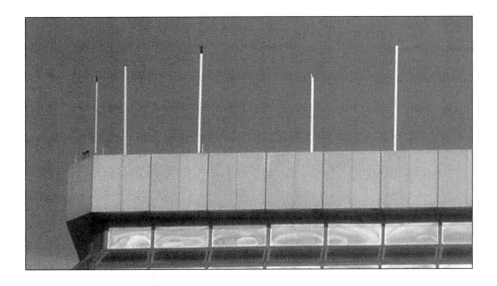

FIGURE 7.1 An antenna site

SAFETY

Because of the height of the antenna, personal safety must be the number one priority of the job. Securing *every* part and tool, so that ***nothing*** drops on any person or property during and after the installation, is a big part of this safety practice.

On a tower or structure, the installers or maintenance technicians should always use safety belts or harnesses. All of the tools should have tethers so that they cannot fall down if the installer loses his or her grip.

ANTENNA TYPES

There are many types of antennas. This section will detail the types of antennas that are used in the radio and wireless business today. We will only be discussing fixed infrastructure antennas; mobile and portable antennas are outside the scope of this book.

FIGURE 7.2 1/4 wave vertical antenna

1/4 Wave Vertical

A 1/4 wave vertical antenna is the most common antenna in use, although it is used in mobile installations more often than in infrastructure systems (see figure 7.2). It is so common because the natural resonant frequency of the antenna is 1/4 wavelength. All of the other antennas are actually variations of the 1/4 wave dipole antenna.

The antenna is actually made of two parts, the main radiating element, and a reflecting ground. On a mobile installation, the body of the car makes the reflecting ground. On an antenna in the air, the reflecting ground must be provided, and is just a rod the same length as the main radiating element, but going in the opposite direction, tied to the shield of the coaxial cable. That's why it is called a dipole antenna.

CHAPTER 7: ANTENNA INSTALLATION

CONCEPT FOR REVIEW:

FREQUENCY VERSUS WAVELENGTH

Have you ever wondered why the antennas on top of police cars are sometimes fifty inches high, sometimes eighteen inches high, and other times only seven or three and a third inches high?

The answer is that antenna length is based upon a formula that relates frequency to the wavelength of the signal. The exact formula is:

Frequency × Wavelength = 300

Here, frequency is in Megahertz, the unit of frequency, wavelength is in meters, and 300 is the speed of light in millions of meters per second.

Because the most efficient radiating length for an antenna is 1/4 of the wavelength, use this formula to determine wavelength, then just divide by four to come up with the length of your antenna.

Example 1

If the frequency is 150 MHz, then the wavelength is two meters, and the antenna would be half a meter, which is approximately nineteen inches.

Example 2

If the frequency is 50 MHz, then the wavelength is six meters, and the antenna would be one and a half meters, which is approximately fifty-eight inches.

Example 3

If the frequency is 450 MHz, then the wavelength is 2/3 meters, and the antenna would be 1/6 meters, which is approximately seven inches.

A 1/4 wave antenna has no gain over a dipole, so it is said to have *unity* gain. There is a fictional antenna called an isotropic radiator. The isotropic radiator is so named because it radiates equally in all directions. The dipole and 1/4 wave antennas will have 2.2 dB gain over this isotropic radiator.

CONCEPT FOR REVIEW:

WHAT ARE DECIBELS AND LOGARITHMS

Logarithms

Numbers can be expressed in two forms. The first is the number itself. The second is how much the number ten (10) must be raised by an exponent to get that number. The examples below will help you understand this:

Example

$$100 = 10^2$$
$$1000 = 10^3$$
$$0.0001 = 10^{-4}$$
$$2 = 10^{0.301}$$
$$3 = 10^{0.4771}$$

Logarithms are useful because they make it easier to express extremely large and extremely small numbers. Logarithms are also useful in expressing relationships between numbers. Any two numbers can be expressed as a ratio, and that ratio can be expressed in dB.

The advent of the scientific calculator has made figuring and using logarithms very easy. Fortunately, you no longer really have to understand the relationship between a number and its logarithm, all you have to be able to do is type in a number in your calculator and push a button. Just make sure you push the right button—there are two types of logarithms, natural and common. The common logarithm is based on powers of ten and is calculated using the key LOG on your calculator. This is different from the natural logarithm, which is calculated by using the LN key on the calculator. You will not be using natural logarithms in the applications we will be discussing.

The gains of an amplifier, a loss of power, audio levels, the power of an earthquake, and many other quantities are expressed in numbers based upon logarithms. Human senses such as light sensitivity, hearing, and pain sensitivity are also measured logarithmically. Sound levels, which are expressed in decibels, are based upon a mathematical formula which is in turn based upon logarithms.

(Continued.)

CHAPTER 7: ANTENNA INSTALLATION

(Continued.)

Decibels

Decibels are the unit of relative measure used throughout the radio world. The following formulas are used for calculating decibels:

Power

$$dB_{Power} = 10 \times LOG\ (P_{Out} / P_{In})$$

$$dB_{Voltage} = 20 \times LOG\ (V_{Out} / V_{In})$$

If you always have the Out divided by the In level, the following will always hold true:

- A positive number represents gain.
- A negative number represents loss.
- If the Input and Output are the same, the gain will be 0 dB.

Effective Radiated Power

A radio transmitter system consists of the transmitter and antenna peripheral equipment. Coaxial line loss, connector loss and antenna gain or loss are predictable, and must be considered by the engineers in the system design. The Effective Radiated Power (ERP) is the amount of power equivalent to that generated by a transmitter placed on the tower with a unity gain antenna. The transmitter power, minus the losses, plus the gains, gives you the ERP.

Power(ERP) = Power(Transmitter) – Power(Losses) + Power(Gains)

dBm

dBm, or "decibels relative to one milliwatt," is a unit of absolute power, where dB is a relative number. All of the values of dBm are based on the definition of 0 dBm as 1 milliwatt. The following chart will help show this:

$$dBm = 10\ LOG\ (P_{out}) / 1mw$$

0 dBm = 1 milliwatt (1 mW)

+ 30 dBm = 1 watt (1 W)

(Continued.)

(Continued.)

+ 60 dBm = 1000 watts (1 kW)

−30 dBm = 1 microwatt (1 uW)

Power Ratios and Decibels

Here are some examples of power ratios and the decibel equivalents.

X2 = 3 dB

X4 = 6 dB

X10 = 10 dB

X100 = 20 dB

X1000 = 30 dB

X 1/2 = −3 dB

X 1/10 = −10 dB

Voltage Ratios and Decibels

Here are some examples of voltage ratios and the decibel equivalents.

X2 = 6 dB

X4 = 12 dB

X10 = 20 dB

X 100 = 40 dB

X 1/2 = −6 dB

X 1/10 = −20 dB

Folded 1/4 Wave Vertical

A folded 1/4 wave antenna is similar to a 1/4 wave antenna, with the exception that the antenna element is tied to both the coaxial cable and the ground (see figure 7.3). The advantage of this antenna is that the ground offers protection from wind-borne static electricity and lightning, as well as providing a wider bandwidth for the operating frequency band.

CHAPTER 7: ANTENNA INSTALLATION

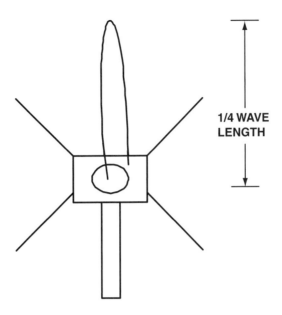

FIGURE 7.3 A folded 1/4 wave antenna

FIGURE 7.4 An omnidirectional 2 element antenna

2 ELEMENT EXPOSED DIPOLE

A 2 element exposed dipole antenna will have a gain of 3.0 dB over a 1/4 wave antenna or a dipole (in free space away from any building, tower, or other antenna). The orientation is not critical as long as the antenna elements are opposing each other (see figure 7.4). This will provide an

FIGURE 7.5 A 2 element exposed antenna with an elliptical pattern

omnidirectional pattern, which means that it radiates in all directions equally. If the antenna elements are both on the same side or only ninety degrees from each other (see figure 7.5), then the orientation is important. This will give a directional pattern, which means that it will favor some directions more than other directions. An elliptical pattern favors two opposite directions and limits the side directions. You will need to contact the system design engineer if the orientation was not spelled out in the installation documentation.

4 Element Exposed Dipole

Similar to the 2 element exposed dipole antenna, the 4 element exposed dipole antenna will have a gain of 6.0 dB over a 1/4 wave antenna or a dipole (in free space away from any building, tower, or other antenna). The orientation is not critical if the antenna elements are opposing each other (see figure 7.6). This will provide an omnidirectional pattern, which means that it radiates in all directions equally. If the antenna elements are both on the same side or only ninety degrees from each other (see figure 7.7), then the orientation is important. This will give a directional pattern, which means that it will favor some directions more than other directions. An elliptical

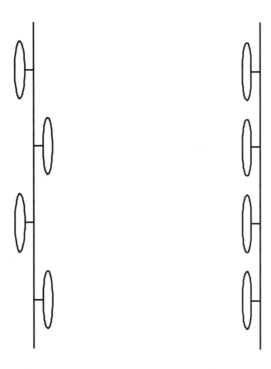

FIGURE 7.6 An omnidirectional 4 element antenna

FIGURE 7.7 4 element exposed antenna with an elliptical pattern

pattern favors two opposite directions and limits the side directions. You will need to contact the system design engineer if the orientation was not spelled out in the installation documentation.

If an elliptical pattern is desired, the gain will be an extra 3.0 dB over the omnidirectional pattern (see figure 7.8).

8 Element Exposed Dipole

Similar to the 4 element exposed dipole antenna, the 8 element exposed dipole antenna will have a gain of 9.0 dB over a 1/4 wave antenna or a dipole (in free space away from any building, tower, or other antenna). The orientation is not critical if the antenna elements are opposing each other (see figure 7.9). This

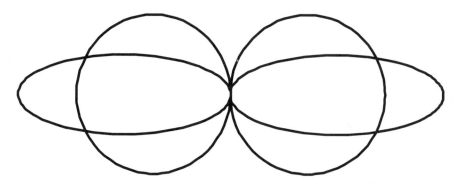

**UNITY GAIN vs. 6 DB GAIN ANTENNA PATTERN
SIDE VIEW**

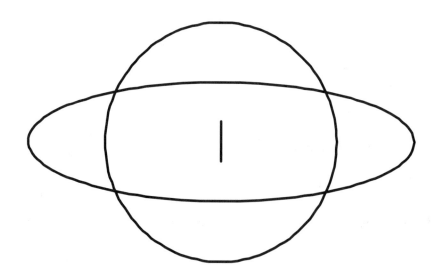

**CIRCULAR vs ELLIPTICAL PATTERN
TOP VIEW**

FIGURE 7.8 Gain comparing unity, omni, and elliptical patterns

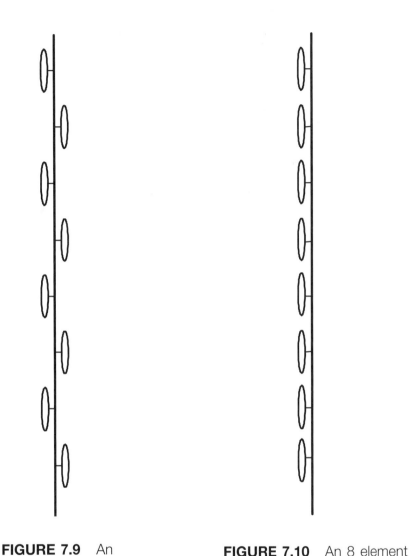

FIGURE 7.9 An omnidirectional 8 element antenna

FIGURE 7.10 An 8 element exposed antenna with an elliptical pattern

will provide an omnidirectional pattern, which means that it radiates in all directions equally. If the antenna elements are both on the same side or only ninety degrees from each other (see figure 7.10), then the orientation is important. This will give a directional pattern, which means that it will favor some directions more than other directions. An elliptical pattern favors two

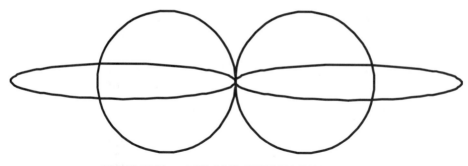

**UNITY GAIN vs. 9 DB GAIN ANTENNA PATTERN
SIDE VIEW**

FIGURE 7.11 Gain comparing unity and elliptical patterns

opposite directions and limits the side directions. You will need to contact the system design engineer if the orientation was not spelled out in the installation documentation.

If an elliptical pattern is desired, the gain will be an extra 3.0 dB over the omnidirectional pattern (see figure 7.11).

Yagi Antenna

A Yagi antenna is also called a beam antenna (see figure 7.12 and figure 7.13). It has high gain (see figure 7.14) and good front-to-back ratio of gain to rejection, but it only works over a small frequency bandwidth.

A beam antenna can be mounted so that the electromagnetic waves are in the vertical plane or the horizontal plane. The industry uses both, so be sure to know which one your site requires before installing one of these beam antennas.

The elements all have names:

- Driven element
- Reflector
- Director
- Director (x)   x = 0 to 18

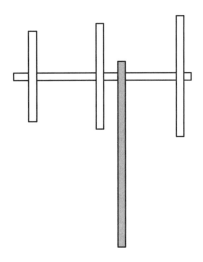

FIGURE 7.12 A diagram of a Yagi antenna

FIGURE 7.13 A photo of three Yagi antennas

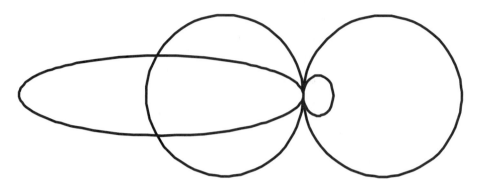

**UNITY GAIN vs. 10 DB GAIN YAGI ANTENNA PATTERN
TOP VIEW**

FIGURE 7.14 The gain pattern of a Yagi antenna

A 3 element Yagi has a gain of 7.0 dB, while a 20 element Yagi has a gain of 20 dB.

Corner Reflector

A corner reflector is a dipole antenna with a metal screen or plate behind it to give a good forward gain (see figure 7.15). The screen should allow **no** signal to be radiated behind the antenna (see figure 7.16).

Log Periodic

A log periodic is a beam antenna, but it has been specially designed to have a **very** broad frequency bandwidth (see figure 7.17). Television antennas are usually log periodics. In exchange for the broad bandwidth, it does give lower gain for the number of elements (see figure 7.18).

Discone

A second kind of very broadband antenna is the Discone (see figure 7.19). It is an ultra-wideband VHF/UHF antenna that is omnidirectional and vertically polarized. These antennas are normally found at airports. Amateur radio operators and the military also use these.

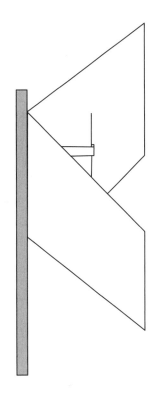

FIGURE 7.15 A corner reflector

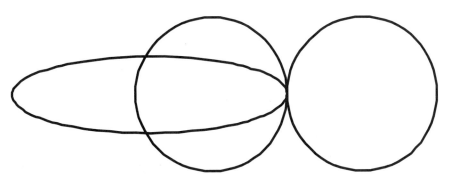

**UNITY GAIN vs. 10 DB GAIN CORNER REFLECTOR ANTENNA PATTERN
TOP VIEW**

FIGURE 7.16 The gain pattern of a corner reflector

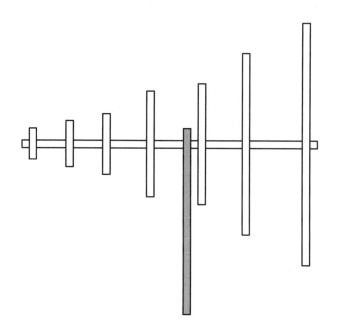

FIGURE 7.17 A log periodic

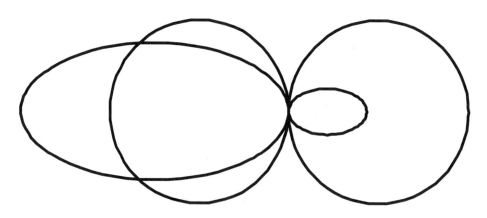

**UNITY GAIN vs. 8 DB GAIN LOG PERIODIC ANTENNA PATTERN
TOP VIEW**

FIGURE 7.18 Gain pattern of a log periodic

CHAPTER 7: ANTENNA INSTALLATION

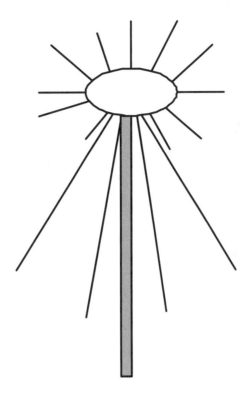

FIGURE 7.19 A Discone antenna

FIBERGLASS COLLINEAR

In some environments, such as near an ocean or where heavy snow and ice will coat an antenna, weather conditions can cause the antenna to corrode or become detuned to the extent that performance will be seriously degraded. To counteract this, antenna manufacturers have developed antennas that are totally enclosed in fiberglass enclosures (see figure 7.20). Some of these antennas can be up to twenty feet in length.

These antennas come in many varieties, with different bands, gains, or patterns, depending upon which one is ordered. The down side of these antennas is that they are much more susceptible to lightning damage, so areas that have numerous lightning strikes should not use these kinds of antennas unless other protective factors are present.

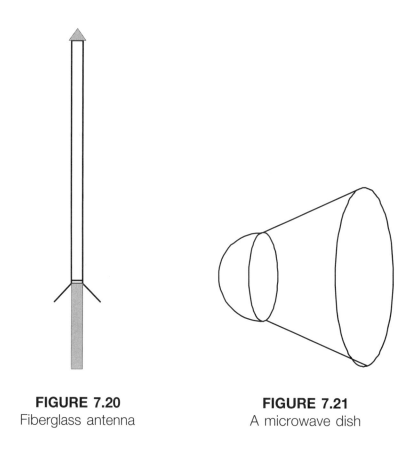

FIGURE 7.20
Fiberglass antenna

FIGURE 7.21
A microwave dish

Microwave Dish Antenna

Microwave dish antennas vary in size from just a few inches to twenty feet in diameter (see figure 7.21). The actual size of the dish is determined by the frequency band and the desired gain.

The radio wave can be horizontally polarized or vertically polarized, so you need to know which direction is appropriate for your site before you mount the dish, and then you'll need to confirm the polarization once the dish is in place. A small dipole inside the dish antenna is the element that must be set.

The beamwidth of a dish antenna is extremely narrow, usually less than one degree, so aiming the dish is a critical part of the installation. The aiming process includes vertical aiming, called elevation, and horizontal aiming,

CHAPTER 7: ANTENNA INSTALLATION

> CONCEPT FOR REVIEW:
>
> ## WHAT IS ANTENNA POLARIZATION?
>
> When you look at the antenna on a car or truck, what you normally see is a rod standing straight up. When you look at an antenna on the roof of a house or attached to the chimney that is used for receiving a television signal, that antenna has its elements in a horizontal plane that is parallel to the ground.
>
> In the case of the vehicular antenna, that antenna is said to be vertically polarized. It will receive signals that are transmitted in the vertical plane over fifty times better than if the signal was coming from a horizontally polarized transmitting antenna.
>
> Television signals are transmitted in the horizontal plane, and the receiving antennas are likewise set to receive in the horizontal plane. Those antennas are said to be horizontally polarized.
>
> Some systems, such as satellite and microwave systems, can function in either plane. As a result, you must know which polarization is required, and you must be sure that you have set the antenna up for that correct plane.

called azimuth. You will need a high-quality magnetic compass and a calibrated level to complete these settings. Modern installers often use waypoints on a handheld GPS unit to determine a general aiming direction.

Because the beamwidth is less than one degree, the mounting must have good stability so that the signal is not lost during heavy winds or storms.

Dish Satellite Antenna

There are many radio systems that use satellites for relaying information to or from the radio to somewhere else (see figure 7.22). In these systems, the dish must be aimed in elevation, by azimuth, and by polarity. Just as with the microwave dish installation, you will need a high-quality magnetic

FIGURE 7.22 Satellite antenna

compass and a calibrated level to complete these settings accurately for a satellite dish installation.

Because the beamwidth is less than one degree, the mounting must have good stability so that the signal is not lost during heavy winds or storms (see figure 7.23).

Panel Antenna

The 800 MHz and higher land mobile radio systems, such as cellular, Specialized Mobile Radio (SMR), Personal Communications Systems (PCS), ISM, Wireless LANs, and iDen™, have developed to the point where sectorized antenna patterns are part of the system design. The manufacturers have met this requirement with panel antennas (see figure 7.24).

The antennas are available for a given radio band, and some have more than one band in the housing. The sector size ranges from forty-five degrees up to 120 degrees. The gain of these antennas varies from unity to seventeen dB of gain.

CHAPTER 7: ANTENNA INSTALLATION

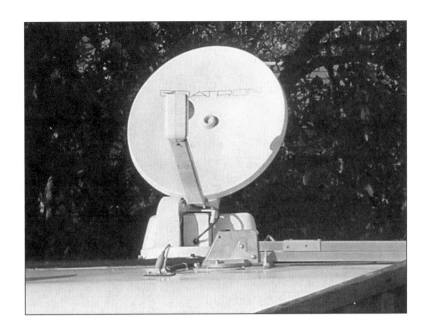

FIGURE 7.23   Another kind of satellite antenna

FIGURE 7.24   Panel antenna

FIGURE 7.25 GPS antenna

Most panel antennas have an adjustment to control the downtilt angle, so that the main lobe of the signal does not overshoot the intended area of coverage. This can be accomplished either by using a mechanical tilting mechanism or by moving the phasing of the elements that the antenna has a downtilt to the pattern.

GPS Antennas

Many sites use Global Positioning Satellite (GPS) receivers to provide very precise time signals. These receivers have an antenna that must also be mounted so that they have an unobstructed view of the sky.

Mountings

There are many types of antenna mountings. Which type you will use is determined by the structure on which the antenna is going to be mounted.

FIGURE 7.26   Round member clamp

If the tower has round members for the vertical supports or horizontal stabilizers, a round member clamp kit is required (see figure 7.26). The size of the member determines the size of the clamp kit.

If the tower's support structure has flat surfaces, then V-bolts and back-support pipes are required for the mounting (see figure 7.27).

If you are mounting the antenna onto a building parapet, then you will probably have to design a special custom mounting. No two sites have the exact same mounting schemes. Some common methods to secure antennas are to use anchor bolts or all-thread through the walls to a custom-made plate or board. Occasionally, you will find a rooftop that has pipes welded into place for the antenna installations.

FIGURE 7.27 Flat member V bolt clamp

Sometimes an antenna is mounted upside down due to space constraints on a tower. If this is the case, be sure that there is a drip hole at the bottom (which is really the top), and that the hole at the top (which is really the bottom) is plugged. If you don't make sure that water can drip out of the antenna mounting pole, it will fill with water and expand and burst during the first hard freeze.

The distance from the antenna to the tower is critical. If the antenna is too close to the tower, it seriously affects the radiation pattern of the antenna (see figure 7.28).

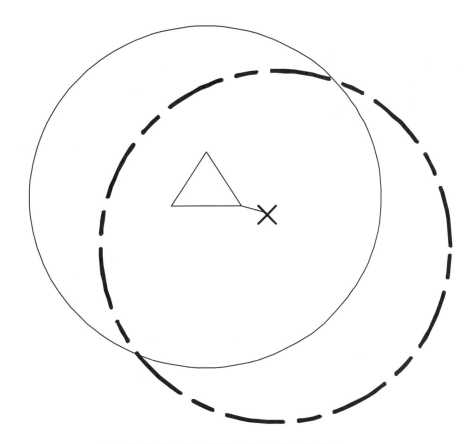

OFFSET RADIATION PATTERN DUE TO TOWER

FIGURE 7.28 The distortion effects of the antenna pattern

INSTALLATION TECHNIQUES

Because the antenna is always outdoors and rust is a major problem at antenna sites, all of the material used in mounting antennas is either made from hot-dipped galvanized iron or from stainless steel. The antennas themselves are usually made of brushed aluminum.

Some of the larger antennas are constructed and shipped from the factory in multiple pieces. When you install these antennas, the physical dimensions

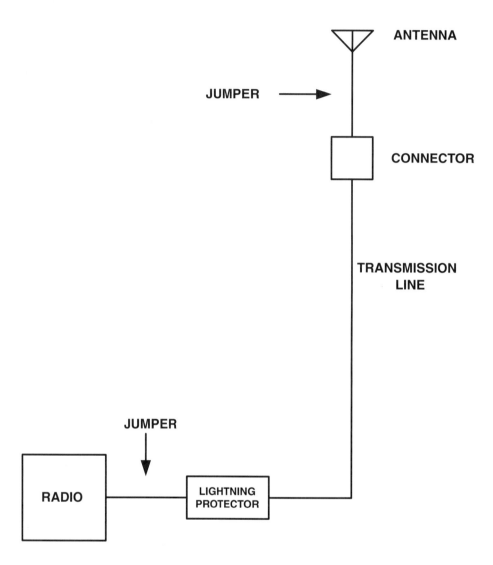

FIGURE 7.29 Jumpers in line

of the elements of the different pieces are critical, and you must be sure to accurately align them during the installation. Also, you will usually use a wiring harness to connect the multiple pieces together to make it into a single antenna with gain. The harness allows the antenna pieces to be phased together, but if the dimensions are off between the elements, your antenna will experience loss instead of gain.

CHAPTER 7: ANTENNA INSTALLATION

> **CONCEPT FOR REVIEW:**
>
> ## WHAT IS GALVANIZATION?
>
> Many of the components used to mount the antennas, as well as the towers themselves, are almost always made of steel. One of the properties of steel is that it rusts. Rust will weaken the tower or mounting hardware over time. To counteract this, a compound that allows tin to be coated over the steel is bonded to the steel components. This stops the rust from forming.
>
> There are two methods for galvanizing steel. The first is a process called hot-dipped galvanization, where the entire piece is submerged in a vat that contains the molten tin. The second process is called cold galvanization, where the tin is painted on using either a spray or dipping process. The hot-dipped process is better.
>
> If the tin is scraped or penetrated in any way, rust will form on the steel below. If a radio signal is in close proximity to the rusted area, severe interference can and will be generated. The rust itself does not cause the interference, but the junction of the rust with another piece of metal will.
>
> Many tower installations require that all steel components be hot dipped galvanized, and that all mounting hardware be made from stainless steel.

Use lock washers on all nuts, and make sure they are tightened until there is no danger of the nut coming loose.

You will use a jumper between the main transmission cable and the antenna in almost every installation (see figure 7.29).

CONNECTOR INSTALLATIONS

The antenna manufacturer determines the type and sex of the RF connector. If at all possible, do not use an adapter. There is a proper series and type of connector made by every manufacturer in the business. Besides, the most popular scheme is (and has been for a while) to use a short (two- to six-foot) jumper between the main transmission line and the antenna. This allows the jumper to act like a fuse in case of a minor lightning hit, so that all you need to replace is the antenna and the jumper, instead of the antenna, the jumper, and the transmission line. The second advantage to using a jumper is that the main transmission line cannot bend into place properly, and the lighter, more flexible jumper fills the bill easily. The last advantage of using a jumper is that you can assemble all of the connectors on the ground and just connect them together once you have them in place on the tower or rooftop. Many companies that use jumpers have the connectors premade at the factory or distributor, so that they do not have to make them on-site.

Once the connectors are installed and in place, the main thing that you must do is weatherproof the connections. You do this by making sure the connectors are tightly secured, then applying a layer of weatherproof electrical tape to the connector and the cables for a few inches near the connector, and finally rewrapping the tape over itself many times. Other sealants are also available, and at least one company, Teracom™, has developed weatherproof connectors. If you are using plain old electrical tape, though, moisture will eventually get in because of a property of chemistry called capillary action. Capillary action explains how liquids, such as water, will have a tendency to creep into extremely tight spaces one molecule at a time. Eventually you end up with a relatively large accumulation of moisture. You will need to apply a coat of something over the tape to prevent the capillary action. This is usually a material called Scotchkote™, which is liquid rubber. Warning: If you get this material on your clothes, it will **not** wash out, as it is waterproof. Amply apply the Scotchkote™ all over the taped connection, let it dry for a half an hour, then tape over the complete connection again. This will make a connection that will last forever, or until the first lightning hit.

INSTALLING CABLES

Most coaxial cables have a companion hoisting grip that is used by the installation company to help pull the cable up the tower or building. This is a stainless steel wire mesh contraption that grips the coaxial cable more tightly as more pressure is put on it, with a loop for attaching the pulling

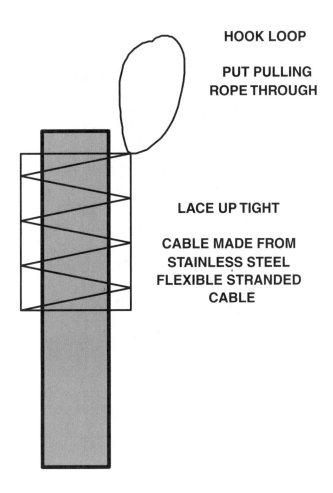

FIGURE 7.30 Hoisting grip

rope to the cable (see figure 7.30). The pressure is distributed along the length of two feet or so of the cable and it does not damage the cable at the gripping points. For cables over 250-300 feet, additional grips are added every 200 feet or so, depending on the size of the cable and the manufacturer's recommendations.

If you are installing a cable that requires multiple hoisting grips, be sure to install them before the cable is raised. Some types of hoisting grips are

FIGURE 7.31   Cables secured using cable clamps

closed, and can only be installed before the cable run has started. In other words, the hoisting grip is a tube, and you'll need the end of a cable to thread through the tube. Other types, called split-type hoisting grips, can be laced-up into place around a cable during the installation.

Note: Be sure that the loop of the hoisting grip is towards the top of the building or the tower.

In a tower installation, the main transmission line should be secured to the tower supports every three to five feet using cable brackets mounted on the tower (see figure 7.31).

Some older towers do not have bracket mounts, and on these towers the transmission lines are secured to the legs of the tower (see figure 7.32). If that

CHAPTER 7: ANTENNA INSTALLATION

FIGURE 7.32 Cables secured to the tower leg

is the case, then you will usually use insulated copper-weld wire or ultraviolet immune plastic tie cable wraps. If you cannot find this copper-weld wire, 10 gauge solid electrical wire will suffice.

> **Caution: Do not overtighten the twist until it deforms the shape of the cable in any way, no matter how slight the deformation is.**

As a precaution to keep lightning out of a site, a grounding kit is normally placed on three strategic points of the transmission line:

- Near the top, close to the end of the transmission line
- Near the bottom, just before the transmission line leaves the tower
- Near the building, just before the transmission line enters the building

Attach the grounding kits to the outside of the transmission line, taking great care not to penetrate the outer metal of the cable. Then completely cover the joint of the kit with the tape supplied in the kit. Secure the tail of the ground lead to the tower so that a lightning spike can go to ground onto the tower leg or building ground bus.

We will discuss this in greater detail in Chapter 9, Grounding.

It is very important that you take great care during the installation of the transmission line so that the cable does not get kinked, bent, or damaged in any way. The shape of the cable sets the impedance, and even a minor dent will cause this to vary from the designed value. A Frequency Domain Reflectometry line sweep will see every kink and sharp bend in the cable.

The cable should always have a height depression, or drip loop, just before it enters the building, so that rain water cannot enter the building through the cable entrance (see figure 7.33).

TESTING

As you can see from the descriptions in this chapter, the antenna, jumper, and transmission line installations are quite complex and time consuming, and the installations are made in such a way that the connections cannot be broken easily. Because of this, each component should be tested individually on the ground before it is installed on the tower. This will save much time later if there is a problem.

There are three main components:

1. Antenna
2. Jumper
3. Transmission line

There are five ways to test the components:

1. Physical inspection
2. DMM

CHAPTER 7: ANTENNA INSTALLATION

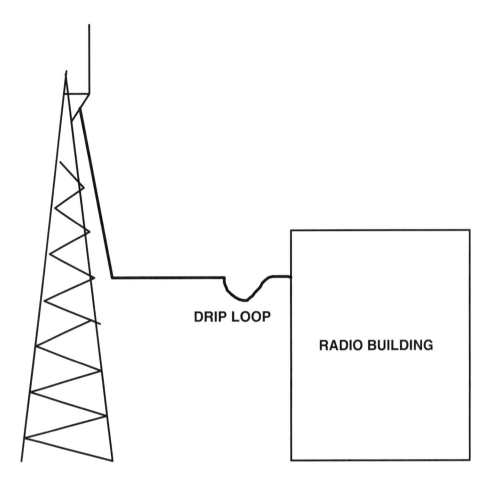

FIGURE 7.33 Drip loop

3. Transmitter power
4. Time Domain Reflectometer
5. Frequency Domain Reflectometer

These methods are described in detail below.

You should perform a thorough physical inspection of every component before the installation is started and after it is completed. Even though there may still be problems that you cannot see, problems that can be visually detected should always warn of a faulty component.

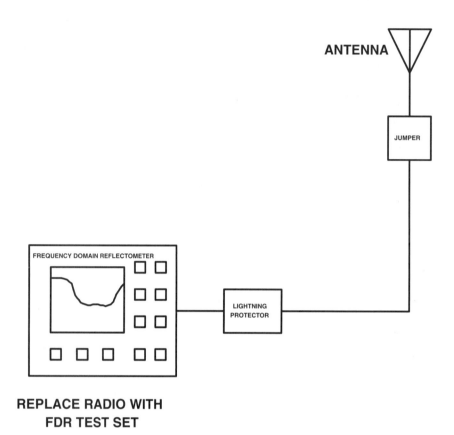

FIGURE 7.34 FDR test setup

Use a DMM to check that there is no short between the center conductor of a coaxial cable and the shield. This resistance value should be infinity, so any resistance that you detect will indicate a problem. Unfortunately, the rest of the system cannot be tested with a DMM.

The problem with using the transmitter to check the system while the antenna is still on the ground is that minor problems will go unnoticed. In addition, you could be exposing yourself and others to excessive radiation if the antenna being tested is close to anybody.

A Time Domain Reflectometer puts a pulse on the transmission line, and waits for a return from either a short circuit or an open circuit. Frequency sensitive items cannot be measured accurately. See Chapter 6, RF Cabling, for detailed information about how a TDR works.

The absolutely perfect instrument for testing antenna systems is a Frequency Domain Reflectometer (FDR) (see figure 7.34). See Chapter 6, RF Cabling, for detailed information about how an FDR works. Many companies require the use of an FDR plot to confirm that the antenna and all of the components of the antenna system are working properly. A return loss on a coax that is lower than −25 dB is considered good. The antenna must return a reading that is lower than −14 dB to be considered working properly. The connectors should also give a return loss that is lower than −25 dB. Even the slightest kink or cable abnormality can be detected with FDR. You can use a standard Vector Network Analyzer (VNA), which is a much larger, much more expensive FDR with more capability, or a special VNA like the Anritsu SiteMaster™ or the Bird Antenna Analyzer for radio sites.

GAS-FILLED TRANSMISSION LINES

Some cables use compressed air or nitrogen to keep moisture out of the lines. The air pump for compressed air is called a dehydrator (see figure 7.35). If there is more than one line, you'll need a multiunit line splitter.

FIGURE 7.35 A dehydrator air distribution system

Summary

The antenna is one of the most important parts of a radio system. The installation is serious business, with no room for error or compromise. Safety is the number one job for a tower installer. There are many types and models of antennas, and it is the job of the system engineer to select the antenna that is most appropriate for a given site. The job of the installer is to ensure that the components are not defective and that they are installed exactly as the engineer has designed the system. Furthermore, all of the pieces must be properly installed to survive bad weather and wind, and be correctly grounded. Finally, the components should be tested before the installation begins and again once the complete system is assembled.

CHAPTER 7: ANTENNA INSTALLATION

QUESTIONS FOR REVIEW

1. What component has the greatest effect on the range of a radio system?

2. Is special insurance required for climbing towers?

3. What is the most important part of an antenna or tower installation?

4. What is the natural frequency of an antenna?

5. What is the gain of a 4 element omnidirectional exposed dipole antenna?

6. Why are jumpers used?

7. What device is used as an aid to pull the coax up a tower?

8. What should be done to keep rain out of the transmitter building?

9. What is capillary action?

10. What instrument fully tests all components and the complete system?

11. What is the device called that pumps air into an air-filled transmission line?

12. Can a kinked cable be detected?

CHAPTER 8

TELEPHONE WIRING

OBJECTIVES

After completing this chapter, you will understand the following concepts:

- Methods for wiring a standard telephone in a radio site
- Methods for setting up a telephone wiring board
- Methods for setting up a main frame
- The definition of a private line (PL) circuit
- Requirements for installing a PL circuit
- Methods for testing a PL circuit

KEY TERMS

- Line
- Station
- Tip
- Ring
- Dialtone
- Demarc

- RJ11
- Station wire
- R66B
- RJ21
- Private line
- T1
- RJ45
- Loopback
- Data circuit

Introduction

A radio site is the place where the "wireless" originates. Making everything work together requires many wires at the wireless site (see figure 8.1). This chapter will explain some of the major points of the wiring process and connectors you'll be using to set up the site's telephony system. This chapter will also discuss connector blocks, standards such as the color codes, telephone boards and Main Distribution Frames, mechanical mountings, and standard telephone wiring. The latter part of this chapter will discuss private line wiring.

Telephone Lines

In order for the radio equipment to work, it is normally interconnected with other equipment at a different site. This can be a dispatcher console, a mobile telephone switching office, a paging terminal, or some other control equipment. Also, in order for installers and maintenance technicians to be able to communicate with these other sites, a standard telephone is normally installed at the site. A multiuser site will have a separate telephone for each client that desires to have telephone contact.

CHAPTER 8: TELEPHONE WIRING

FIGURE 8.1   A telephone board at a site

At most sites, the telephone company will bring in a multipair terminal block for a common interconnection point. This terminal block is called the demarcation, or demarc, point and it divides the wiring responsibility between the phone company and the wireless site. The telephone company is resposible for any problems with the wiring and equipment on the side going back to the Central Office from the demarc point. The individual who is leasing the line is responsible for the wiring and equipment on the side going to the rest of the site. The demarc block itself is the responsibility of the telephone company.

The telephone company keeps track of the lines by numbers attached to the cable and cable pair. They also use a circuit ID number. You can find these numbers on the RJ21 blocks, where the telephone lines terminate in most sites, and on the demarc block. All of these numbers are important, and if you do need the assistance of the telephone company, it will need all three numbers when you report that there is a problem on the line.

The telephone company normally has the demarc on one of three types of blocks. The first is a block that is normally made of fifty pairs of terminals, each of which uses a hex nut to hold the wire in place. The second is a fifty pair block that uses a slotted or Phillips pan-head screw to hold the wire in place. The last type that is used by the telephone company is a twenty-five pair or greater (in twenty-five pair increments) block that uses punch-down terminals on the block.

No matter which type of physical connector the telephone company provides, the line is interconnected from the connector to the rest of the site. Some sites use a telephone board, while other sites use a main frame.

TELEPHONE STATIONS

A standard telephone line is called a Plain Old Telephone Service, or POTS, line. People, both inside and outside the telephone company, use this term. A single pair line from the telephone company to the telephone instrument, called a station, is all that is required to make the station work. The directory number is made up of the area code and the seven digit number we usually think of as the phone number. If you have any problems with this line, you will use the directory number as the circuit ID when you call the telephone company for assistance.

TELEPHONE BOARD

A telephone board is usually a four foot by six foot by 3/4 inch or larger plywood board on which all of the telephone blocks are mounted. It is usually filled with R66B blocks or RJ21 blocks. An RJ21 is an R66B with an Amphenol™ twenty-five pair connector on one or both sides (see figure 8.2). A punch-down tool is used to attach the wires to the block. All of the telephone lines and equipment lines are terminated on the telephone board, and jumpers are used to tie the equipment to the lines.

In order to keep the wires looking nice and straight, most telephone boards either have a spindle or use bridal rings to keep the wires grouped together in a straight pattern.

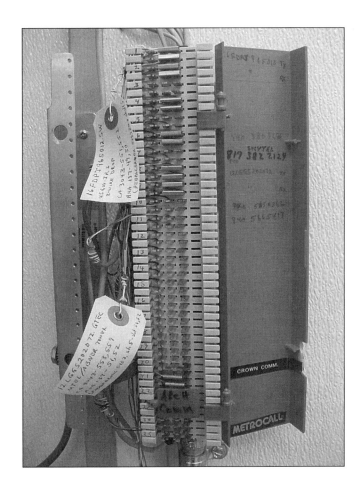

FIGURE 8.2 RJ21 block

Main Frame

When there are more than 200 lines terminating on a telephone board, the board can become very crowded and require more room than most facilities have available to accommodate the phone systems. Telephone companies have thousands of lines to deal with, and the solution that they have come up with is the use of a Main Distribution Frame (MDF). The MDF will accommodate thousands of terminations and cross connections in a compact, orderly fashion.

The vertical side of an MDF ties to the telephone lines going in or out of the building. The horizontal side goes to the equipment terminations. All of the wires from the lines or the equipment are terminated on WireWrap™ blocks, which can hold 100 pairs or more per block in a four-by-six foot space. The interconnection between the lines and equipment is via jumper wire that is installed in an orderly, neat arrangement.

WIRING

Most telephone stations are wired from the telephone board or from the MDF using a four-conductor cable called station wire. This is usually a cream colored cable with a red, green, yellow, and black wire inside of the cable sheath. One or two telephone lines can be wired using this cable. The first, or primary, line uses the green lead for tip and the red lead for ring of line one. If you are using a second line, then the yellow lead is the tip and the black is the ring for line two.

If you are planning to run the station wire along a wall or vertical flat surface, you will usually staple it into place using round staples, and you will have to take care to make sure that the staple does *not* cut through the insulation and short out the conductors inside the cable sheath.

When working with this station wire, take extreme care that nobody steps on the cable, and that no heavy object is placed on top of the cable, as the insulation is designed to be broken or stripped with very little effort.

CONNECTORS

The standard connector for telephone station instruments is the RJ11 connector (see figure 8.3). This connector can accomodate one or two pairs, and if you use a special version called an RJ14, even three pairs. The first line always connects to leads 3 and 4. The second line always connects to leads 2 and 5. If you are using the RJ14 and a third line is on the same connector, then leads 1 and 6 are used.

CHAPTER 8: TELEPHONE WIRING

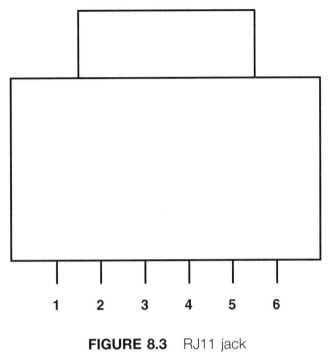

FIGURE 8.3 RJ11 jack

JUMPERS

The jumper wire is usually 22, 24, or 26 gauge in size, and is solid wire. It will work with punch-down, screw, or nut type of connectors.

If you nick the wire while stripping the lead, you should cut the lead at the nick and restrip the lead again. The wire becomes very weak and easy to break at any nick. If you have to reuse a wire that has been connected using a screw or nut, the wire is usually nicked by the mechanical connection and you should cut and strip it for any reuse.

When using the punch-down tool, the yellow side of the tool usually has the cutting edge. People in the industry call this "sunny side down"—maybe that will help you remember which side to use. If you use the tool correctly, the wire will be seated into the punch block and cut on the first push of the tool. You can only put one wire into each terminal of the punch block.

Most jumper wire has a two-color scheme. The normal telephone wire color code uses the following colors:

Color	Tracer
Blue	White
Orange	Red
Green	Black
Brown	Yellow
Slate	Violet

The lead that is predominantly the tracer hue is the tip, while the predominant color hue is the ring. On the R66B or RJ21 blocks, the tip is lead twenty-six on the Amphenol™ connector, but is the top lead on the block. An Amphenol™ connector is a twenty-five pair connector used throughout the telephone industry. It is named after the Amphenol Corporation, which originally manufacturered these connectors. The ring is lead 1 on the Amphenol™ connector, but is the second lead from the top. The remaining forty-eight leads alternate from tip to ring until the bottom of the block, which is lead 25 (ring).

LIGHTNING PROTECTION

Most telephone companies provide a lightning protector before the demarc block (see figure 8.4). Sometimes, their demarc block is the lightning protector. The most common type of protector is a solid-state Varistor type that provides a path to ground on either the tip or the ring when the voltage exceeds a preset limit, usually 200 volts. A second type of protector is a carbon fuse that opens the circuit when a lightning strike gets into the line. A third type is a gas tube that conducts when the voltage exceeds a few hundred volts. The second and third type described here sometimes short to ground after a lightning strike, and after one of these has been hit by lightning, anyone using that line will hear a hum in the audio.

CHAPTER 8: TELEPHONE WIRING

CONCEPT FOR REVIEW:

TELEPHONE CABLE COLOR CODE

We included this table in Chapter 3, Equipment Wiring, but it is listed again here so that you will not have to flip back and forth when wiring up a site that uses telephone switchboard cabling.

TELEPHONE CABLE COLOR CODE

Pair	Ring		Tip	
1	Blue	White	White	Blue
2	Orange	White	White	Orange
3	Green	White	White	Green
4	Brown	White	White	Brown
5	Slate	White	White	Slate
6	Blue	Red	Red	Blue
7	Orange	Red	Red	Orange
8	Green	Red	Red	Green
9	Brown	Red	Red	Brown
10	Slate	Red	Red	Slate
11	Blue	Black	Black	Blue
12	Orange	Black	Black	Orange
13	Green	Black	Black	Green
14	Brown	Black	Black	Brown
15	Slate	Black	Black	Slate
16	Blue	Yellow	Yellow	Blue
17	Orange	Yellow	Yellow	Orange
18	Green	Yellow	Yellow	Green
19	Brown	Yellow	Yellow	Brown
20	Slate	Yellow	Yellow	Slate
21	Blue	Violet	Violet	Blue
22	Orange	Violet	Violet	Orange
23	Green	Violet	Violet	Green
24	Brown	Violet	Violet	Brown
25	Slate	Violet	Violet	Slate

FIGURE 8.4 Telco lightning protector

PRIVATE LINE CIRCUITS

I introduced POTS lines in the early part of this chapter. The rest of this chapter concentrates on private lines, including T1 lines. There are more private lines in use at radio and wireless sites than there are POTS lines. POTS lines are part of the Public Switched Telephone Network (PSTN), while private lines are dedicated, direct lines between two or more points. You can't call anywhere on a private line except the location at the other end of the line, so there are no dial tones or phone numbers with these systems. In addition, there are usually test facilities built into the lines so that they can be tested without requiring telephone company personnel to be on-site. For these dedicated systems, the wiring needs to be of the highest standard because the radio site is totally dependent on these lines.

Chapter 8: Telephone Wiring

Private Telephone Lines

POTS lines use one pair of wires to operate, while a private line always has two pairs coming from the telephone office. One of the lines is used for transmitting from the site to the Central Office, while the other is used for receiving from the Central Office.

Another characteristic of a private line is that the demarc point must be placed after a device called a loopback set. With this loopback set, the telephone company can verify if a line is working properly, and remotely control the setting of audio levels from a central operations center. The loopback set requires 120 VAC, so a standard electrical outlet must be available for each loopback set at the site.

A special type of private line that is sold as a product by the telephone company is a T1 line. This is a specially formatted data circuit that can carry twenty-four voice channels, two television channels, eight ISDN channels, or some combination of these. If you need more than eight lines from the telephone company, it is usually cheaper to bring the lines over one T1 line rather than bring in nine or more separate lines.

Private Line Wiring

The wiring from the demarc point is usually made with jumper cable to the punch-down or WireWrap™ block of the equipment. There are usually two pairs associated with each circuit. The first pair transmits audio from the equipment to the telephone Central Office, and the second receives audio from the Central Office to the equipment.

The transmit pair is called the tip and ring. The receive pair is called the T1 and R1 pair.

Private Line Connectors

Most equipment that is interconnected with other sites uses RJ21 or Amphenol™ connectors, or RJ45 connectors. An RJ45 looks like an overgrown RJ11

Wiring for Wireless Sites

CONCEPT FOR REVIEW:

WHAT IS A T1?

When you pick up your telephone at your house, and talk to another person, you are most likely using a single pair of wired cables between your house and the telephone company Central Office or sub-office.

In order to conserve on the number of lines needed between Central Offices, the telephone company allows twenty-four conversations to be carried on a single pair of specially formatted wires called a T1 line by using a time division multiplexing scheme.

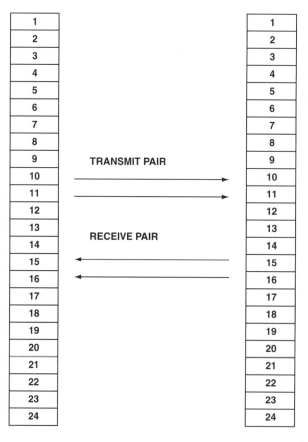

(Continued.)

(Continued.)

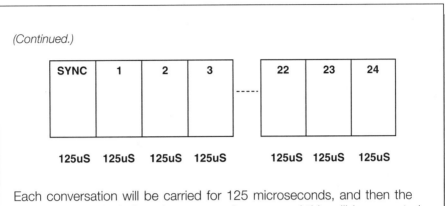

Each conversation will be carried for 125 microseconds, and then the next will be carried, then the next, and so on, and this will be repeated 8,000 times each second. This will allow twenty-four conversations to be transported over the two pairs, one in each direction.

All lines between telephone Central Offices are using T1 lines that utilize this scheme, or a derivation of this scheme.

connector, with eight conductor contacts (see figure 8.5). The pin-out between transmit and receive conductors varies greatly. There is no real standard for the pin-out on a RJ45 connector. There is a standard, however, for the interconnection of the RJ45 with a T1 connection.

Private Line Jumpers

The jumpers for a private line are the same as they are for a POTS line, but there is a second pair for each circuit. Usually, there are two sets of colors, so that it is easy to separate the transmit pair from the receive pair.

Private Line Testing

Private lines must have the transmit and receive levels set in the equipment and on the lines themselves. The equipment levels are set with a transmission line test set. The telephone line levels are set by the Network Operations Center (NOC) at the telephone company, and this can be done without a telephone company representative present at the wireless site.

FIGURE 8.5 RJ45 connector

If there is no signal, this is called a loss of continuity. The telephone company can test its lines via the loopback set at the site (see figure 8.6). When the loopback set is enabled into the loopback mode, the receive pair from the Central Office is crossed back to the transmit pair back to the Central Office. If they test okay, then the problem is at the site and must be analyzed by a technician at the site.

A transmission test set is capable of sending and receiving various tone frequencies at precise levels. It is imperative to have one available to test with the telephone company personnel in order to confirm the wiring and levels at the site.

Summary

A functional telephone system is vital to pretty much every business, and the wireless industry is no exception. The wiring from the telephone company comes into a site, is demarcated by the telephone company and

CHAPTER 8: TELEPHONE WIRING

FIGURE 8.6 Loopback set

then cross-connected to the equipment. If this equipment is a telephone station, the line goes straight to the station instrument. If a lot of lines are brought in from the telephone company, a Main Distributing Frame (MDF) will accommodate hundreds or thousands of lines in a relatively small space. If an MDF is not required, then all of the cross connects are made at a telephone board. The most common type of interconnection block is an RJ21 or R66B block. Jumpers are used between the blocks and are kept straight and neat using bridal rings or spindles. The telephone station itself will use the two middle positions of a RJ11 block, which are leads 3 and 4 (the red and green leads) of the block. The workmanship of the interconnection of any telephone jumper or cable should be of the utmost quality. Finally, lightning can get into the telephone cable, so a good lightning protector must always be present at every site. Ususally, the telephone company provides this, but if they haven't, then the installer should notify the system engineer and ensure that one is present. Private lines are circuits leased from the telephone company that connect one point to another or that connect multiple points. There is **no** dial tone associated with private lines. The circuit has two pairs, a transmit pair and a receive pair. The connectors are usually RJ45 or Amphenol™ connectors. The jumpers used to interconnect the line with the equipment is made of two pairs, using different color schemes for transmit and receive pairs. The testing is done locally with a transmission test set, or remotely by the telephone company technicians at the NOC using a loopback set which is located at the wireless site.

QUESTIONS FOR REVIEW

1. What are the leads called in a telephone circuit?
2. What is the standard connector that a station uses called?
3. What side of a punch-down tool has the cutting edge?
4. What does POTS mean?
5. What does MDF mean?
6. What is a demarc?
7. What is the tip and ring pair direction in a private line?
8. What is the T1 and R1 pair direction from the Central Office?
9. What does a loopback set do?
10. How many conductors are in an RJ45 connector?
11. How many voice channels can be carried by a standard T1 circuit?

CHAPTER 9

GROUNDING

OBJECTIVES

After completing this chapter, you will understand the following concepts:

- The definition of a ground
- Reasons why a is ground necessary
- Methods for attaching a ground to the equipment
- Definition of a halo ground
- Definition of a ground ring
- Methods for attaching a ground to the tower

KEY TERMS

- Ground
- Grounding
- Ground ring
- Halo ground
- Compression lug
- Split bolt

Introduction

A good ground is an important part of a radio site for the following reasons:

- It helps reduce damage in case of a lightning hit
- It allows all of the equipment to be at a common potential
- It prevents any chance of personnel being electrocuted if a short occurs to a frame or cabinet from the AC wiring
- It reduces the chances of stray RF getting into the desired signals
- It is required by most manufacturers

What Is a Ground?

At a radio site, a ground is a set of large cables going from every rack, piece of equipment, frame, doorway, and other metal components, including the tower itself, to a common bus which goes to the main ground point of the site. A good ground ensures that the potential difference will be the same, even during a lightning strike. In order to accomplish this, the connections to the ground wire must be of extremely low resistance.

Tower Ground

The tower must have a good ground. This is accomplished by placing a good conductor into the earth around the tower and making sure that the earth itself has good conductivity. If the soil around the tower does not have good conductivity, the conductivity can be improved by putting ionizing minerals and water into the ground when the tower base is constructed, and using special chemicals inside the main ground point. There are a few companies making kits and supplying the ionizing minerals to accomplish this.

The ground must be attached to the tower leg or legs. This is done with #8 gauge wire or larger. The lead is normally attached through a cadweld connection. This will be covered in more detail later in this chapter, in the section devoted to cadwelds.

This same ground post will have a cable that goes to the ground bus inside the building.

OUTSIDE RING GROUND

In order to improve lightning dissipation during a strike, bury a bare metal cable around the tower at a distance of ten to fifteen feet away from the tower, with at least four radials extending from the central ground post to this ring. This cable is called the ring ground, although it is not really necessary to make it circular.

From the ring, bury a cable that ties to everything metal in the vicinity of the tower, including fences, buildings, guy posts, and anything else made of metal (see figure 9.1).

INSIDE RING GROUND

Inside the site, near the top of the room, install a bare copper cable around the perimeter of the room. Run cables from this ring to all the metal components and frames inside the site.

GROUND BUS

In order to make it easier to ground all of the equipment in a room and all of the cables coming into the room, install a ground bus near the cable entrance plate (see figure 9.2). Install a lightning protector for each cable near the cable entrance and tie each lightning protector to this ground bus.

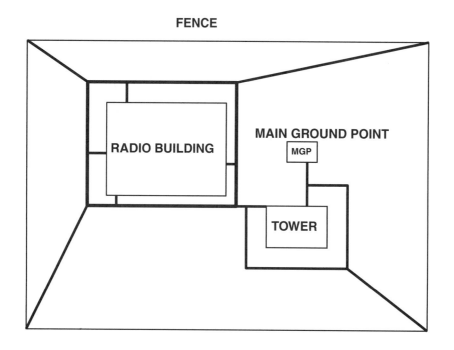

FIGURE 9.1 A diagram of an outside ring ground

FIGURE 9.2 An inside ground bus

FIGURE 9.3 An outside ground bus

Attach a second ground bus to the outside of the building near the cable entrance plate for the grounding kits to which the coaxial cables are attached (see figure 9.3).

EQUIPMENT GROUNDING

You will need to tie the frame of each piece of equipment and of every metal device or item in the room to the ground ring with #6 gauge cable (see figure 9.4). Most sites use stranded copper wire with green insulation. Bare metal is required to make a good electrical connection so you will need to strip all the paint from the area around the connection point. Use dual hole lugs to provide an extremely low resistance connection.

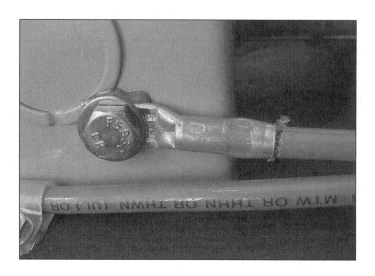

FIGURE 9.4 An equipment grounding example

Coaxial Cable Grounding

You will normally ground coaxial cable in three places as it comes from the antenna into the building:

1. Near the end of the cable close to the antenna

2. Near the bottom of the tower before the cable bends to go into the building

3. Near the building

In each case, you will strip the outer jacket away from the coaxial cable, being careful not to cut into the outer copper conductor, and attach the grounding kit to the coaxial cable. Be sure and get the proper grounding kit to match the specific cable you are using (see figure 9.5). Once you make the connection, seal up the connection using the tape and rubber wrap that is included in the grounding kit. For the two grounds on the tower, attach the pigtail to the tower leg using a clamp, or to a bolt in the tower structure. For the ground that is close to the building, attach the pigtail to the outside ground bus. In addition, ground the lightning protector right at the lower termination connector of the coaxial transmission line.

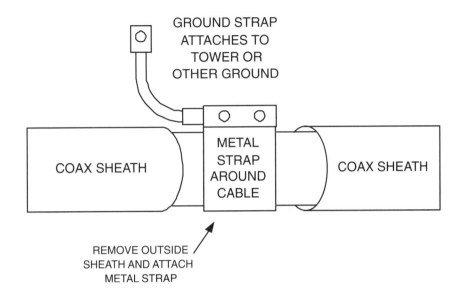

GROUNDING STRAP KIT

FIGURE 9.5 A coaxial grounding kit

Be sure to not have any sharp bends or turns in the pigtail. Remember that lightning often travels at very high frequencies and will not pass sharp turns. Lighting typically follows straight lines or very gentle curves.

Grounding Other Equipment

In addition to the normal equipment frames and cabinets, other possible site components which must be grounded to the ground bus include the following:

- Load center
- Metal door frame
- Metal door
- Spare equipment cabinet

- Air conditioner
- Space heater
- Metal desk
- Metal workbench

GROUND RODS

The ground rods should be copper-clad steel rods with a minimum diameter of 5/8" of an inch and a length of eight feet. There should be at least four ground rods attached to the tower ground ring and at least four ground rods attached to the building ground ring. The ground rods should be no more than fifteen feet from each other, and if you need more than four ground rods to make sure that they are not spaced more than fifteen feet apart, be sure to use more. Precisely uniform spacing is not crucial, but it is important to avoid big gaps in your ground ring.

CABLES

To construct the ground rings around the building and the tower, you should use solid #2 AWG or larger copper wire that has been tinned (see figure 9.6). You can join these wires together using the exothermic welding process. This is called cadwelding. You can also use compression-type crimps that have a twelve-ton or better crimp.

The halo ground, which is a ground wire that circles the room inside the site near the ceiling, should also be a #2 AWG solid or stranded cable (see figure 9.7). Stranded cable is easier to form and work with, but solid cable will work too.

The cable between the inside ground ring and the outside ground ring should also be of #2 AWG.

You should use #6 AWG or larger diameter for any cables that you run between the equipment or any metal piece inside the building and the inside halo ground.

CHAPTER 9: GROUNDING

FIGURE 9.6 A tower ground

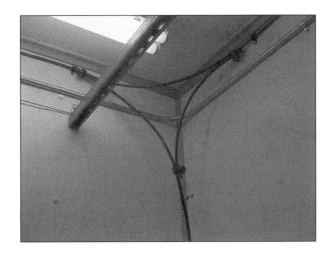

FIGURE 9.7 An inside halo ground

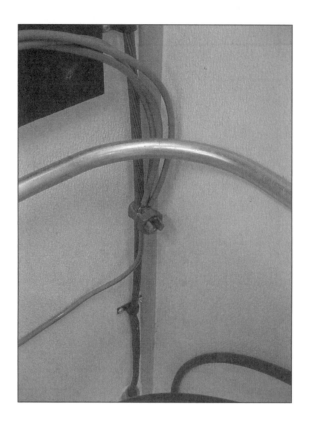

FIGURE 9.8 A gradual bend in a corner

In all cases, the ground cables should have *no* sharp bends. Lightning will jump off the cable rather than take a sharp bend. Use a gradual bend at places where a bend or turn is necessary (see figure 9.8).

Some companies allow aluminum cables to be used instead of copper cables. If you are using aluminum cables, you must use cabling that is two gauge sizes greater than the recommendations I've given you for copper cabling. If you use both copper and aluminum, you'll need to use a bimetallic transition conductor or connector because you cannot have aluminum touching copper. If you do end up with aluminum connecting with copper, it will set up a chemical reaction called electrolysis, eventually creating a high resistance joint that will make the ground ineffective.

CONNECTORS

You should ensure that there is a good bond at all the connectors between the cables and the lugs, and between the lugs and the cabinets or other cables and rings. Make sure to remove any paint at the contact points. Use an anti-oxidant compound such as NO-OX™ in the lugs and at the point of contact. In addition, use star washers to increase the contact point and to ensure that the connection is tight. Some companies require that two-hole lugs be used instead of single hole lugs.

CADWELDS

Cadwelds are connections made by placing a clay-like material, like plastic explosives, onto the outside surface of the tower leg with the cable that is to be welded to the leg. The compound is then ignited with a cigarette lighter. It burns at 3000 degrees, melting the cable and the tower leg together. This weld will allow the installer to attach any cable to a steel tower leg (see figure 9.9). There are many safety precautions that need to be taken when using this material, and it should only be used by people with training and experience with it.

> **Note:** Only an experienced person should use this material, as it is very dangerous.
>
> **IMPORTANT NOTE:** You cannot drill a hole into a tower leg, as this will weaken the leg and it will no longer be able to support the rest of the tower above the hole.

OTHER GROUNDS

All metal parts must be grounded, including fences and gates (see figure 9.10). It is very important that ***everything*** in the site be at the same potential. Failure to follow this rule will cause a potential difference with the first nearby lightning strike, and you'll be repairing damage (or replacing equipment!) that you could have prevented.

FIGURE 9.9 A cadweld on a tower leg

FIGURE 9.10 A gate grounding

Testing

The grounding between any two cabinets or components can be tested with a Digital Multimeter (DMM). The main ground point and the soil around it can be tested with a ground continuity tester called a megger. The megger will measure the conductivity of the soil; see the instructions the came with your megger box for the acceptable pass point.

Summary

The ground is a very important part of a radio site. Every piece of equipment and every metal object in the site and external to the site must be at the same potential, even during a lightning strike. To accomplish this, the wiring, cables, and installation standards for the ground must be of the highest quality, and all procedures must be followed carefully.

QUESTIONS FOR REVIEW

1. How many ground rods should you use around a tower?

2. How can you improve the conductivity of the soil around the site?

3. What gauge wire should you use for the outside ground ring?

4. Why should there never be sharp bends in the grounding cables?

5. What size wire should you use for the inside ground ring?

6. What is a cadweld?

7. Can you drill into the tower leg to secure the ground cable to the tower?

8. Name at least three items other than equipment that must be grounded at a site.

9. At how many different points from the tower to the building must the coaxial cable be grounded?

10. What is the name of the point that all of the grounds are tied back to?

CHAPTER 10

LIGHTNING PROTECTION

OBJECTIVES

After completing this chapter, you will understand the following concepts:

- The various elements of lightning protection systems
- Transmission line protection
- Antenna protection
- AC power protection
- Telephone line protection

KEY TERMS

- Lightning
- Ground
- Spark gap
- Varistor
- Shorted stub

Introduction

Since lightning gets into the equipment via lines and cables that are external to the equipment, the way to prevent this from happening is to put lightning protection on each line that is external to the site. These lines include the antenna and transmission lines, the AC power lines, and the telephone lines. In each case, the lightning protector must be tied to the ground via a good conductor.

CONCEPT FOR REVIEW:

WHAT IS LIGHTNING?

Lightning is an occurrence of nature where charged particles of one polarity build up in magnitude and then conduct to the opposite polarity. This usually occurs during thunderstorms, when the moisture in the air provides a conduction path for the charge. Lightning is usually classified in one of two ways:

- Cloud to cloud

- Cloud to ground

In cloud-to-cloud lightning, you see brilliant flashes going between clouds, and little damage is done on the ground. In cloud-to-ground lightning, a multimillion volt, 200,000 Ampere flash will come out of the clouds and zap anything in its path on the ground.

Even though state-of-the-art lightning control cannot control when and where lightning will hit, it does allow the energy to be dissipated with a minimum amount of damage to nearby buildings and equipment. Lightning protection works by directing the path that the lightning follows. In addition, the majority of lightning hits are small fingers coming off of a main strike. A major bolt of lightning usually destroys everything in its path—no lightning protection system will save your site from a direct hit. But the small fingers of lightning can induce a high voltage into electronic equipment, and current state-of-the-art lightning protectors can protect your equipment from these surges.

CHAPTER 10: LIGHTNING PROTECTION

WHY DO WE NEED LIGHTNING PROTECTION?

No lightning protector will protect a site against a direct lightning hit. The voltage and current of a direct hit will destroy most of the site. Fortunately, the vast majority of lightning strikes are not direct hits, but are only manageable secondary hits, against which the site can be protected. In addition, the probability of a direct hit to a site is small compared to the chances that a hit somewhere else will send lightning to your site via the cabling that is external to the site. This is why lightning protectors work to protect the equipment and the site. A site that does not have this lightning protection will be much more vulnerable than one that is properly constructed.

CABLES

As we pointed out in the chapter on grounding, the ground cables must be of a large enough gauge to be able to withstand these indirect lightning hits and assist in limiting the voltage during the hit. For this reason, the ground must meet the standards described in Chapter 9, Grounding. All of the lightning protectors depend upon the ground cable to carry the lightning back to the earth without doing any damage to the equipment or the site.

CONNECTORS

All of the lightning protectors have some provision that allows the protector to be attached from a cable back to the ground at the site (see figure 10.1). Usually, this is a hole or lug to which the cable is attached. If the joint is not tinned, or if the connector is made of a metal that is different from the cable, then some provision must be made to prevent electrolytic corrosion which will eventually cause the connection to become a high-resistance joint. The manufacturing standards that I referred to in Chapter 1 provide more details about preventing this kind of corrosion.

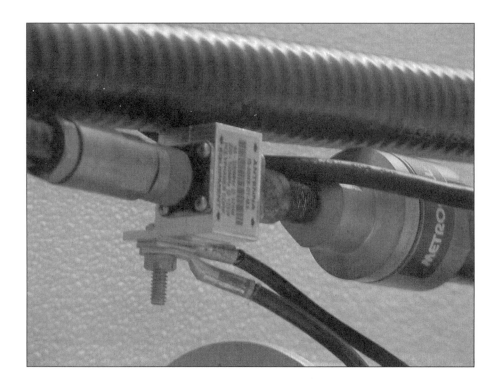

FIGURE 10.1 A cable tied to a lightning protector

AC Protection

The AC power line is a common way for lightning to get into a site. When lightning moves along a power line, you might see some combination of the following three results.

First, a very high voltage spike can occur on the power line (see figure 10.2). A spike is extremely short in duration. Without protection, a spike will burn out many of the electronic components at the site.

The second way that lightning affects a site is a surge in the AC voltage. A surge is longer in duration than a spike (see figure 10.3). Without protection, a surge will burn out many of the components at a site.

CHAPTER 10: LIGHTNING PROTECTION

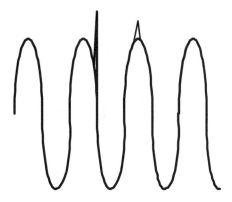

FIGURE 10.2   An AC voltage spike

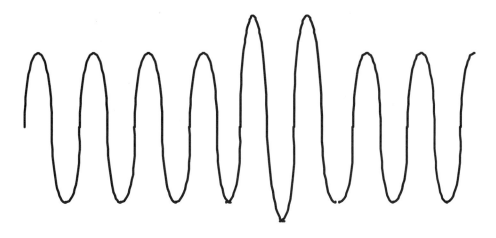

FIGURE 10.3   An AC voltage surge

The third way that lightning effects a site is to cause the total disruption of incoming AC power. You should have a UPS or battery backup system and a backup generator to protect your site against this problem.

A good AC spike and surge arrester will correct the first two problems. The closer the arrester can keep the voltage to normal, the better off your site will

FIGURE 10.4 A ferro-resonant transformer

be in the event of a strike. Many cheap arresters let the voltage exceed the damage point before they react, so make sure that you use quality equipment. Rather than put an arrester or surge suppressor on each piece of equipment at the site, most sites have the surge suppressors wired into the load center or make the suppressor an integral part of the load center.

A ferro-resonant transformer will act as a voltage equalizer to prevent surges and to also give the spike arresters time to react before the lightning can do any damage to a site (see figure 10.4). The transformer must have a capacity large enough to handle the total current of the equipment it is designed to protect.

> **CONCEPT FOR REVIEW:**
>
> ## WHAT IS A VARISTOR
>
> A Varistor is a semiconductor device similar to a zener diode. It acts like a short circuit once the trigger voltage spike is established, and then when the high voltage spike goes back to normal, it goes back to a non-conducting state.
>
> Typically, a Varistor, also called a Metal Oxide Varistor (MOV) will be found on each lead of a power wiring cable or telephone cable, with one side tied to the power or telephone lead, and the other side tied to ground. Since these lines are found in pairs, there are always two MOVs per circuit.
>
> When a spike does come down the line, the MOV will fire within a few nano-seconds, and the spike will be dissipated in the MOV, then, once the voltage spike is gone, the MOV goes back to a passive state of acting like an insulator.
>
> As the MOV absorbs hits, it eventually will short circuit out. Because of this, a fuse or circuit breaker should always be in line between the source and the MOV.

TELEPHONE LINE PROTECTION

Lightning can enter a site through the telephone lines. The telephone company usually provides a spark gap or Varistor type of lightning protector that will short out the line when the voltage exceeds a hundred volts or so. This is sufficient in most cases to eliminate any damage to the telephone line or any equipment tied to these lines. This is the same for POTS lines or Private Lines.

The older telephone company protectors were a carbon button that acted like a fuse whenever a lightning strike caused the voltage to exceed a certain amount (see figure 10.5). You might see these in some of the older sites, and

Wiring for Wireless Sites

FIGURE 10.5 The telephone company's lightning protector

in most homes. This type of protection requires time to react, and may not be fast enough to eliminate all spikes, or protect all equipment, especially delicate electronics such as modems.

Many people do not trust the telephone company to provide a good lightning protector on their lines, so they also use their own lightning protector (see figure 10.6).

Antenna Protection

Some antennas are more prone to lightning damage than others. An exposed element grounded dipole antenna is immune to all but major strikes (see figure 10.7). Fiberglass antennas are vulnerable, and one mounted on the top of a

CHAPTER 10: LIGHTNING PROTECTION

FIGURE 10.6 A telephone line protector

FIGURE 10.7
An exposed element antenna

FIGURE 10.8 A fiberglass radome antenna

tower is the most vulnerable (see figure 10.8). The drawback to exposed element antennas is that they do not work very well when there is ice or snow on the elements. Also, in coastal areas where there is salt water in the air, exposed element antennas will corrode. For this reason, many engineers will use the fiberglass antenna and take a chance on lightning, rather than have to put up with very poor performance during the winter months.

COAXIAL CABLE PROTECTION

Whether the lightning hits the tower or the antenna, some of it will travel down the transmission line to get into the building or site. To prevent this,

CHAPTER 10: LIGHTNING PROTECTION

FIGURE 10.9 A coaxial cable grounding kit

there is a standard that requires you to cut away the antenna coaxial cable outer jacket in three places and then tie the outside shield of the coax to ground.

The first place where you should do this is at the top of the run of coax near the antenna (see figure 10.9). The second place is near the bottom to the coax just before it makes the turn to leave the tower. The final place is just before the cable enters the site. In all three places, take care to make sure that there are no sharp bends in the ground wire and that the ground pigtail goes to a tower leg or to a good ground in a direct fashion.

RF Equipment Protection

In order to protect the RF equipment at a site, use a high-quality RF lightning protector. The RF signal does have to pass through this device, so you want a device that causes little signal attenuation, while still providing lightning protection.

There are three major classifications of RF lightning protectors:

- Spark gap
- Fuse
- Shorted stub

The spark gap type of lightning protector contains an inert gas, such as argon, and when the voltage exceeds a few hundred volts, the gas breaks down and allows the lightning to go to ground via the gap arrester (see figure 10.10). You can still have problems, and once a hit has occurred, there is sometimes a carbon path that interferes with the RF signal. If that is the case, you must replace the lightning protector. In addition, sometimes the lightning gets through anyway and still damages the RF equipment.

The fuse type of lightning arrester has a thin conductor to allow the RF to pass through. A lightning hit, however, with its large current, will blow the fuse and break the circuit between the antenna and the RF equipment. The lightning then takes the path along the frame of the lightning protector, which is always connected to a good ground via a large cable.

The shorted stub lightning protector is a relatively new type of lightning protector (see figure 10.11). It is a DC short between the center conductor and the outside frame, but in the frequency range of operation, a shorted stub looks like an open at a 1/4 wavelength of the operating frequency. As a result, the shorted stub lightning protector is frequency band specific in its allowable operating range.

CHAPTER 10: LIGHTNING PROTECTION

FIGURE 10.10   A spark gap arrester

FIGURE 10.11   A shorted stub lightning protector

> CONCEPT FOR REVIEW:
>
> **HOW DOES A SHORTED STUB WORK?**
>
> There is a natural wavelength that corresponds to each frequency in the world of radio. For more information, see the Concept For Review on Frequency vs. Wavelength in Chapter 7.
>
> One of the principles of radio propagation in a medium is that the impedance flips every quarter wavelength. This dimension will depend on the frequency in operation.
>
> If a transmission line was shorted at the end of a 1/4 wavelength stub, then this would look like an open at the meeting point of the stub. If the transmission line was open at the end of the 1/4 wave stub, it would look like a short at the meeting point of the stub.
>
> By having the line shorted at the stub, all signals would be shunted to ground, including lightning, except that signal on the 1/4 wave operating frequency. This phenomenon works very well and the newer lightning protectors use this principle.
>
> Open stub   Shorted stub

Summary

Lightning can enter a site on the transmission line, coaxial cable center, AC power line, or telephone line. Lightning protectors are used to keep lightning out of the sites by protecting each of these access points. There are many brands and types of lightning protectors, and the engineers are charged with providing the correct one for each site.

Questions for Review

1. Name the four access points that must be protected?
2. What is the effect of a direct lightning strike?
3. What does the site lightning protection protect against?
4. How many ground kits should be on each transmission line?
5. What is the best type of antenna to use in areas that are extremely lightning-prone?
6. What is the common lead on all lightning protectors?
7. What are the three types of RF lightning protectors?
8. How does a spark gap lightning protector work?
9. How does a fuse type lightning protector work?
10. How does a shorted stub lightning protector work?

CHAPTER 11

MISCELLANEOUS WIRING AND CABLING

OBJECTIVES

This chapter will explain which low voltage circuits are found at a site and provide you with the information you'll need to install and maintain these circuits, including the connections you'll need to make. The circuits introduced here include the following:

- Data circuits
- Alarm circuits
- Sensor circuits

In addition, this chapter will explain the requirements for cables and connectors, and identify those times when local wiring licenses may be required.

This chapter is also written to help you understand the Fiber-Optic systems found at wireless or radio sites. Some of the main points that we will cover include the following:

- Fundamentals of Fiber-Optic systems
- Safety precautions
- Cable types

- Protection requirements
- Connector types
- Installation requirements
- Testing requirements

KEY TERMS

- RS232C
- Normally open
- Normally closed
- Single-mode
- Multimode
- IR
- nm

INTRODUCTION

The construction and ongoing operation of a site requires many circuits and peripheral cables that connect to low voltage circuits. These include data circuits, peripheral data equipment, alarm circuits, and sensor circuits. The cables and connectors are sometimes critical, so we will discuss these areas in detail later in this chapter.

In addition, many wireless and radio sites have Fiber-Optic systems that link equipment and controllers. This chapter will discuss many of the items that an installer or maintenance technician will encounter if Fiber Optics are used at the site.

CHAPTER 11: MISCELLANEOUS WIRING AND CABLING

DATA CIRCUITS

In addition to the data circuits that go to telephone lines outside of the site, many sites have data lines that run to various peripheral equipment components. These include printers, CRTs, computers, modems, and display units.

The primary standard that data devices use for the interconnections between different pieces of equipment is called RS232. This is a standard that the Electronics Industries Association (EIA), a group of electronic equipment manufacturing companies, came up with in the early 1970s that defines the voltages and pin out connections between different pieces of data sending and data receiving equipment. Almost all of the equipment that uses serial data communications uses this standard.

A minimum of three wires, and as many as twelve, can be used to transport the data from one device to the next. These data cables are relatively immune to noise and RF interference, but the original standard placed a fifty-foot maximum on the length of the cable between devices.

In order to accommodate longer runs of cable, RS232C can be converted into current loop signals, then the length can be extended to many thousands of feet.

Many sites require that the installer run these cables over to the desired location across overhead racks (see figure 11.1) or under raised floors. Rarely will it be acceptable to leave cables just lying on the floor or running across aisles where someone could trip, possibly snagging equipment and pulling it onto the floor.

ALARMS AND SENSORS

Most radio sites are unattended; therefore, it is very common for site mangers or operators to install remote monitoring systems. Some of conditions the sensors monitor include the following:

FIGURE 11.1 Data cable running to overhead rack

- Perimeter security
- High temperature
- Low temperature
- Water level
- AC power status
- Generator function
- Air conditioner function
- Presence of smoke
- Presence of fire
- Equipment on-line
- Battery voltage

CHAPTER 11: MISCELLANEOUS WIRING AND CABLING

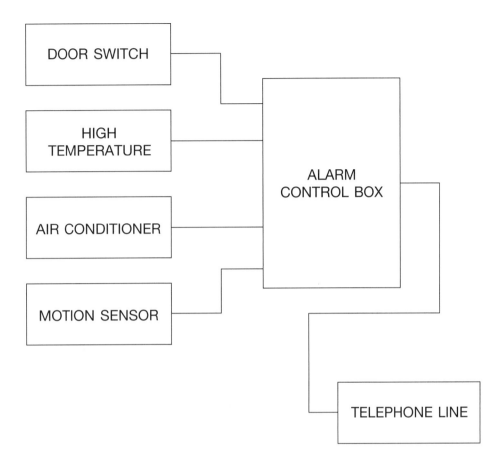

FIGURE 11.2 A typical alarm system

- Fuse integrity
- Circuit breaker integrity

Many of the sites have a central alarm controller (see figure 11.2). The input from the sensors can be one of the following: a normally open circuit, a normally closed circuit, or a voltage level.

A normally open (N.O.) circuit means that the sensor is open when the sensor is in the off or negative mode. When a problem occurs, the sensor causes the circuit to complete, current flows, and the controller sees the

215

closed condition as an alarm for the condition that is being monitored. The problem with an N.O. sensor is that a cut wire or faulty sensor will not be discovered until a physical inspection or test of the sensor is made. For this reason, N.O. sensors are not normally used.

In a normally closed (N.C.) circuit there is current flowing through the sensor contacts at all times, and the interruption of the current causes the alarm. A faulty cable or sensor will also trigger an alarm, so this is the desired mode for most sensors.

A voltage sensor alarm is usually connected to a special set point alarm sensor. When the voltage either rises above or drops below the set point, the alarm is triggered.

Cables and Connectors

Regardless of the type of sensor that is used, the current to power the alarms is usually very low, so even the smallest gauge wire will work satisfactorily. Many sites use telephone station wire to connect the sensors to the central alarm monitoring system (see figure 11.3).

Most of the sensors use screw terminals, so all you need is the stripped wire or a spade lug to make the interconnection.

A few systems will use special connectors, but those will usually be supplied with the system.

Low Voltage Wiring License

Many states in the United States, including California and Nevada, require that installers obtain a special low voltage installers license in order to wire up these low voltage circuits. In most situations, there is no high voltage or danger involved with working on these circuits, but the law overrides any other order. If you are working in one the states that require it, make sure you have the license in-hand to perform these installations.

CHAPTER 11: MISCELLANEOUS WIRING AND CABLING

FIGURE 11.3 Alarm box

CABLE TESTING

Almost every low voltage cable or circuit can be tested with a DMM. Test the cables themselves at the proper points for continuity. Also test the equipment for the correct potentials at the proper points.

FIBER OPTIC EQUIPMENT REQUIREMENTS

The engineers for the site will work with the manufacturing company to determine what Fiber-Optic systems are to be installed at a site. If Fiber Optics are required, then the installers will need to put in the Fiber Optics and perform the added testing that goes with the installation. Once the system is installed, the maintenance team takes over, and they will have to perform the same tests again.

> **CONCEPT FOR REVIEW:**
>
> ## FIBER OPTIC SAFETY PRECAUTIONS
>
> The wavelengths of the light that the Fiber Optic (FO) communication systems use are outside the visible spectrum of light. As such, a dark fiber, which means that there is no signal present, looks just like an active one at first glance.
>
> The problem that arises is that the energy from a live FO cable can cause blindness if it gets into your eyes. You cannot see the emission from the fiber, but you will discover the effect after you have no vision.
>
> For this reason, always treat **every** Fiber-Optic cable as though it were turned on and just waiting to destroy your eyesight.
>
> Always keep the end covered when it is not terminated into a jack.
>
> The glass fibers inside the cable are also very small, so always use safety glasses or goggles whenever you handle FO cable.
>
> Finally, **never** look directly into the end of a Fiber-Optic cable unless you are 1000% sure that the signal is off.

There are two main bands or wavelengths that are presently used in Fiber-Optic systems. These are 1310 nano-meters (nm) and 1550 nm. There are also two types of Fiber-Optic cable—single-mode and multimode. For short runs and jumpers, it does not matter which type you use, as long as the mode of the cable matches the connectors and terminations to the equipment. For long runs of cable, single-mode is the preferred type of cable.

FIBER OPTIC CONNECTOR TYPES

Many types of connectors are used with today's Fiber-Optic systems (see figure 11.4, figure 11.5, and figure 11.6).

CHAPTER 11: Miscellaneous Wiring and Cabling

FIGURE 11.4 ST connector

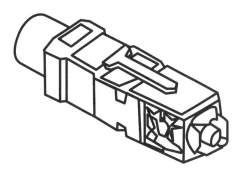

FIGURE 11.5 SC connector

FIGURE 11.6 FC connector

There are both single-mode and multimode versions of each of these connector types; the cables can be either single-mode or multimode as well. In addition, there are other types of connectors, and many require special tools to make the connection correctly. Make sure you have the proper type of connector and installation tools for the job.

If you have to install the connectors onto the cables on-site, you will need special tools, and there are special procedures you'll need to follow. If you are not familiar with these tools and procedures, or if you do not have the tools and testers, do not attempt to attach the connectors. Get outside help from someone who has the proper resources.

CONCEPT FOR REVIEW:

WHAT IS THE DIFFERENCE BETWEEN SINGLE-MODE AND MULTIMODE FIBER-OPTIC CABLES?

When you select a Fiber-Optic cable for use in a communications system, you have to make a decision right away: what mode of cable do you use?

The size of the glass conductor inside a Fiber-Optic cable comes in two sizes, called single mode and multimode. The outside dimension of the cable is the same, regardless of whether the conductor inside is single-mode or multimode.

Multimode cable allows the light waves to bounce around, off the side walls of the internal conductor. This is because the size of the conductor is 62.5 micro-meters (um) in diameter. This bouncing around allows the light waves to arrive at different times, due to the extra time the reflection waves take in traveling and bouncing between the walls of the conductor. The extra room inside of the conductor allows the wavelength of the light to spread.

Single-mode cable has the same outside diameter, but the inside conductor is only 9 um in diameter. The small diameter of the conductor prevents the light beam from being able to bounce off the internal walls of the conductor. This ensures that that the light waves all arrive at the same time.

(Continued.)

(Continued.)

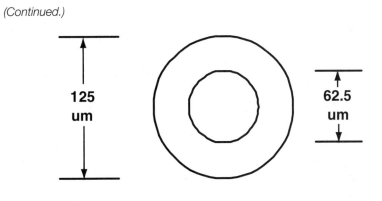

Multimode Fiber-Optic cable

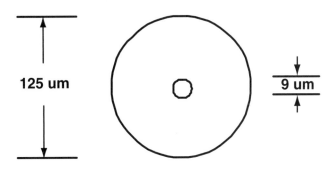

Single-mode Fiber-Optic cable

The maximum speed of the signal will have to be reduced if you are using multimode cable for a distance of more than a couple of miles in length.

If you are installing a short patch cable, a multimode cable will work in either type of system, but a single-mode patch cable will only work in a single-mode system. If your system uses multimode cable, you won't be able to use single-mode cables even for short patches.

FIGURE 11.7 The ribbed plastic conduit protects the Fiber-Optic cables

PROTECTING FIBER CABLES

Fiber-Optic cables are somewhat fragile, so it is standard today that all cables should be run through a brightly colored plastic pipe (see figure 11.7). There should be no sharp bends or kinks in the cables. Most engineers will order a pipe big enough so that the cable can be fed through the pipe even with the connectors attached. This way the cables can be ordered with the connector terminations from the factory.

FIBER OPTIC CLEANING

Whether you make your own connection terminations or use a premade termination cable, you need to confirm that the end is clean and that it will not contaminate the equipment connector. You must use special cleaning swabs designed for just this purpose every time you place a connector into service.

CHAPTER 11: MISCELLANEOUS WIRING AND CABLING

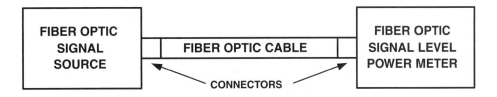

FIGURE 11.8 Fiber-Optic cable test setup

If you do not have these cleaning swabs and you plan to install a Fiber-Optic cable, get them before the connection is put in place. A clean connection is not optional.

FIBER OPTIC TESTING

Every time that a cable or jumper is used, whether premade or made on-site, it should be tested to ensure that it is working properly.

In order to test a Fiber-Optic cable, you should use the same wavelength (1310 nm or 1550 nm) and the same mode (single- or multimode) as the operating signal. A good test setup will include a source and a signal receiver (see figure 11.8). The signal attenuation (loss) of the cable will be given in decibels on this receiver, and you should record that number for future reference. In addition, you should do a visual inspection with an X100 or better microscope to confirm that there are no fibers or other debris in the connector end.

SUMMARY

Low voltage circuits such as data, alarm, and sensor circuits are found at most sites. The interconnection of the circuits should be done in a neat, orderly fashion. The standard for data circuits is usually RS232, while the alarms and sensors are usually just a single pair of leads between a sensor and a monitor box. Some states require a low voltage license, and if that is the case in your state you will need to make sure the installer is licensed before he or she starts wiring these circuits. Fiber-Optic systems are coming into vogue as a way to interconnect equipment at a site, and to interconnect sites with the main controller sites. There are two main wavelengths in use today, two types of cable mode, and many types of connectors. Because of the combination of connector type, mode, and installation scheme there area myriad of connectors on the marketplace. All Fiber-Optic cables need to be cleaned, inspected, and tested before being placed into service.

CHAPTER 11: MISCELLANEOUS WIRING AND CABLING

QUESTIONS FOR REVIEW

1. What does N.O. mean?
2. What does N.C. mean?
3. Can an alarm be transmitted if a wire breaks on an N.O. sensor circuit?
4. Will a broken wire be detectable on an N.O alarm circuit?
5. Will a broken wire be detectable on an N.C. alarm circuit?
6. Is a low voltage installers license required in every state of the U.S.?
7. What does nm stand for?
8. What kind of Fiber-Optic cable is used for long runs?
9. Name three types of Fiber Optic connectors.
10. What type of protection should a Fiber-Optic cable have?
11. What should be done before installing any Fiber-Optic cable?

CHAPTER 12

EMERGING TECHNOLOGIES

OBJECTIVES

This chapter will provide you with an overview of newly developed and emerging technologies. It is beyond the scope of this book to provide detailed specifications for these, but you are encouraged to investigate those that are of interest.

INTRODUCTION

The time to plan for the technologies of tomorrow is today. Wireless technologies are changing rapidly as more and more people demand smaller devices with more robust capabilities. Internet applications such as email are now being provided via cellular and PCS telephones to millions of people. Text messaging is available via PCS telephones, pagers, and other devices.

Telematics is an entirely new realm of communications dedicated to providing communication links between host computer systems and a myriad of devices such as automobiles, long-haul trucks, containers, and even small packages during shipment. A goal of this technology is to provide location information and telemetry to a host system for tracking purposes.

The common theme among all of these new technologies is networking. Devices being able to communicate seamlessly with other devices is the key to success in a rapidly changing technological environment. The processes required for two devices to share data and communicate intelligently are commonly described in terms of layers. The reference standard for these layers is the OSI Seven-Layer Model. While a detailed description of this model is beyond the scope of this book, interested readers can research this topic independently.

Two noteworthy layers of the data communications model are the physical and the network layers. Common standards for these have been adopted by most manufacturers and systems developers. Ethernet, as a physical layer interface, and TCP/IP, as a network layer protocol, are practically ubiquitous in networked environments. The deployment of systems that follow these standards has made data communications much easier.

Networking in the Metropolitan Environment

One wireless Metropolitan Area Network (MAN) familiar to everyone is the existing cellular telephone network.

Since the middle 1980s, 800 MHz cellular radio systems have been available to use around the United States. These early systems were analog radio systems that offered basic mobile telephone service. As time went by and technology advanced, the systems were upgraded and digital modes such as Time Division Multiple Access (TDMA) and Code Division Multiple Access (CDMA) were added into the 800 MHz spectrum.

Personal Communications Services

In the early 1990s, the Federal Communications Commission added the frequency range of 1840 MHz to 1990 MHz for use by Personal Communications Services (PCS). These new frequencies and system operators became the competition for the cellular customers on the 800 MHz systems.

Systems that provide PCS are commonly, although mistakenly, also referred to as cellular. While at this time the services offered to the public are basically the same as those offered by cellular systems, the underlying technologies differ significantly. Right from the beginning the PCS systems were digital. In addition, the systems were designed for low power handheld radio telephones, where the original cellular systems were designed for the more powerful mobile radios with highly efficient antennas mounted on cars and trucks. This meant that there had to be many more base station locations for the PCS systems. Also, the much higher frequency band that PCS uses requires better standards for the installations.

As these technologies proliferate, expand in geographical area, and evolve in terms of services being offered, the need for well-managed and maintained radio sites increases dramatically. This can be best observed in the monopole PCS radio sites that have been constructed over the past few years. These small sites are now commonly located in both commercial and residential areas and are no longer considered the eyesore that they once were.

While these sites differ in their construction from traditional building penthouse radio sites, their function and their basic components are the same. Similar also, is the need to ensure that installation standards are followed and that all work, especially site wiring, is done in a professional manner.

A wireless MAN, as implemented by many of the PCS service providers, is made up of many discrete coverage cells, each having its own radio site. These sites are typically interconnected by multiple T1 telephone lines, as well as being connected to a common network operating center located somewhere in the area.

Third Generation PCS: 3G

There is no single definition or specific technical standard for Third Generation (3G) Personal Communication Services. Many manufacturers and service providers have similar but somewhat differing visions regarding future service offerings. Despite their difference, all refer to their particular vision as 3G. Likewise, upgrades to existing systems and enhancements to carriers' offerings are now being referred to as 1.5G, 2G, or 2.5G. These also are creative definitions in lieu of an objective technical standard.

Definitions notwithstanding, the evolution of PCS systems and the MANs that support them from a voice telephone call only application to an integrated voice and data service is common to everyone's vision.

The seamless integration of wired and wireless technologies that will support Internet applications, web browsing, streaming video, text messaging, and voice communications is a good general description of the promise of 3G.

Networking in the Enterprise Environment

A data communications network within a single facility, building, or common space is commonly referred to as a Local Area Network (LAN). The network that connects multiple LANs together is commonly referred to as a Wide Area Network (WAN).

In a large facility with many PC users, multiple applications, or extremely high bandwidth requirements, users may be segregated into small groups. In this circumstance, the network connecting the members of each group is a LAN, and the network connecting the groups together is a WAN.

The technology used in LANs and WANs is dependent upon the applications being supported and the systems used to run these applications. In today's environment, the most common applications are deployed to users on desktop PCs and may be presented in web browsers or other communications programs.

Networking between PCs and among groups of PCs is the most common deployment of LAN/WAN technology today. As previously mentioned, the standards of Ethernet and TCP/IP support this type of networking and these have evolved to provide robust, scalable connectivity.

One of the most interesting new technologies to gain acceptance in the enterprise environment is the Wireless Local Area Network (Wireless LAN). This technology uses radio links to provide Ethernet connectivity to users' computers instead of fixed wiring. A desktop PC, laptop, or Personal Data

Appliance (PDA) can be equipped with a Wireless LAN adapter and the user will have the same connectivity as if he or she were still "plugged in" to the network. In regards to this technology, the most exciting element is that it frees the user from the traditional stationary desktop PC and enables movement within the enterprise while maintaining connectivity to the network.

The reference standards for Wireless LANs are the IEEE 802.11 and 802.11b specifications. These documents define the physical, electrical, and radio link characteristics of the most commonly deployed Wireless LAN technology. Interoperability between devices is ensured by testing and certification from an independent group; the Wireless Ethernet Compatibility Alliance (WECA). Devices with the Wi-Fi™* certification from WECA meet stringent standards for seamless operation within an 802.11b network. Details regarding the 802.11b specifications or the WECA certification process can be found at the WECA web site, www.wi-fi.org.

NETWORKING FOR THE INDIVIDUAL

A relatively new concept is the Personal Area Network (PAN). Emerging technologies exist to add "intelligence" and networking capabilities to a wide array of devices and equipment. Small electronic devices that require minimal power and that are relatively inexpensive can legitimately be thought of as wearable computers. Many of these already exist in the form of two-way pagers, PCS-enabled PDAs, and data-enabled PCS telephones.

PAN AND PCS

Today's devices, however, are not well-integrated and most only support applications within a community of interest defined by a service provider. The promise of PAN technology is to enable communications between devices within a small physical space, independent of any service provider.

Presently, your PCS telephone on your service provider's network enables you to place and receive calls through its network. Your two-way pager, or

---

* Wi-Fi is a registered trademark owned by the Wireless Ethernet Compatibility Alliance

PCS-enabled PDA may provide data communication, email, or even micro-browsing capabilities. Again, however, these functions are based on wireless communication links between the device and the service provider's network. PAN technology, as presently conceived, will add intelligence and capabilities to these devices, enabling them to become "aware" of each other and to share information more directly.

Bluetooth™*

Bluetooth is a networking technology currently being developed for deployment in the PAN and LAN environments. The Bluetooth specification defines a short-range wireless network that enables device-to-device and device-to-network connectivity via a small transceiver located in the work, home, or commercial environment. Detailed information regarding Bluetooth can be found at www.bluetooth.com.

Bluetooth enjoys a broad acceptance in the development arena and many manufacturers are adding Bluetooth capabilities to existing devices. Depending on the specific application, Bluetooth functionality may or may not be intended to replace existing wired or wireless network connections. A Bluetooth-enabled PC may communicate with a Bluetooth-enabled printer, and in this case the wired connection between them would be eliminated. A Bluetooth-enabled PCS telephone, however, would continue to use its existing wireless link to the service provider's network for its primary functions of sending and receiving calls. Upon entering a Bluetooth-enabled workspace, however, additional functionality would be available. A Bluetooth-enabled device will "register" upon entering the workspace. As a part of the specification, all such enabled and registered devices become aware of each other's presence.

While in the Bluetooth-enabled workspace, it may be more efficient for a user to receive data transmissions via a wired network up to the nearest Bluetooth transceiver than to receive it via a service provider's Wide Area Network. The advantage of this would be that the carrier's network is not

---

* Bluetooth is a trademark owned by Bluetooth SIG, Inc.

used for the transmission, realizing spectral efficiency and potential cost savings. One of the goals in deploying Bluetooth technology is to allow host systems to make intelligent decisions regarding the location of individuals and the most efficient method of communicating with them via Bluetooth-enabled devices.

Another goal of Bluetooth is the ability for a user with an enabled device to personally register with the network. In this scenario, Personal Area Networking is achieved. A Bluetooth-enabled device, programmed with user specific identity information enters a Bluetooth enabled workspace. A host system recognizes the user and logs him or her as being at that particular location.

Bluetooth technology has been developed in extremely miniaturized form factors that require minimal power and that are relatively inexpensive. These characteristics make Bluetooth an exciting technology with much promise for future applications. Among the concepts that exist for Bluetooth deployment are enabled and intelligent credit cards, vending machines that can automatically recognize a user and, via a Bluetooth device, authorize a purchase, and even Bluetooth-enabled clothes that can register the wearer upon entering a Bluetooth-enabled store.

700 MHz Systems

In order to alleviate the overcrowding of the radio spectrum, the FCC has set aside the 700 MHz radio band for public safety (police, fire, EMS, etc.) and other dispatch type radio systems (power companies, business, etc.). This frequency band is presently occupied by UHF television broadcast stations. The FCC has told these 700 MHz UHF broadcasters that they must move to a different frequency by 2006, and at the same time, the FCC has told the public safety radio users to start preparing to move to 700 MHz in 2006.

None of the radio equipment that public safety and dispatch systems are using now will work on this new band. This will create a demand for a complete new infrastructure of radio systems in 2006.

RFID

Instead of using bar codes to identify items in stores, companies have developed a technology called Radio Frequency Identification (RFID). This technology uses a radio transponder that uniquely identifies each item using a radio signal from a strategically located transmitter/receiver combination. This technology will be widespread in a few years.

Summary

The wireless world has gone from just voice communications over a radio channel to voice, data, video, email, telematics, and other uses. The systems range in size from metropolitan area networks to personal area networks. Frequencies have expanded to include those that were never thought of two decades ago. The applications of wireless will be a part of everyday life for everyone.

APPENDIX A

ANSWERS TO ODD-NUMBERED QUESTIONS

The only answers you will find here are for the odd-numbered questions. Visit the Delmar Thomson Learning electronics technology website, www.electronictech.com, for the answers to all of the questions.

CHAPTER 1

1. **Name at least two things you need to take into consideration to ensure a safe site.**
   - Proper AC and DC power installation
   - Proper Radio Frequency (RF) installation and safety factors

3. **What are the three types of sites?**
   - Buildings at the base of a tower
   - Tall building penthouse or rooftop
   - Pad at the base of a tower

5. **Name three specification agencies that specify requirements at sites.**
   - NEC – National Electric Code
   - NFPA – National Fire Protection Agency
   - UL – Underwriter's Laboratories

7. **What is a ground?**

   A common electrical connection with the same electrical potential as the earth. A good ground is a requirement for lightning protection and the safety of personnel.

9. **What types of cabling and wiring are found at a site?**

   - AC power
   - DC power
   - Antenna(s)
   - Inter-bay wiring
   - Alarms and sensors
   - Interconnection
   - Telephones
   - Data circuits
   - Controllers

11. **Who is responsible for a site?**

    The site licensee. The installer may have contractual obligations, but the licensee is ultimately responsible.

13. **What is Halon?**

    A fire suppression gas used in equipment rooms. Halon has recently begun to be replaced by FE36.

15. **Is air conditioning required at a site during cold weather?**

    Yes! Communications equipment generates a tremendous amount of heat. Air conditioning is usually required—even in winter.

17. **Why is site documentation important?**

    Documentation is important to prove that you have the proper permits and licenses to construct and operate your site. Documenting all aspects of the wiring and power systems will also reduce the amount of time required to troubleshoot a site or conduct preventive maintenance.

APPENDIX A: ANSWERS TO ODD-NUMBERED QUESTIONS

CHAPTER 2

1. **Name at least three hazards that can exist at a radio site.**
   - Equipment falling on personnel constructing the site
   - Sharp edges of metal pieces that can cut personnel
   - Improper or unsafe use of construction tools
   - Floors that cannot withstand the weight of equipment
   - Obstructions in walkways

3. **What determines the layout of a multi-bay equipment system?**

   The manufacturer determines the layout of a multi-bay system.

5. **How close to level are you required to make the equipment?**

   The bubble on the level must be within the lines.

7. **What precautions should you exercise when cutting racks?**

   Protect the cut edges, and clean up filings.

9. **What precautions must you use if stud guns are used for anchoring equipment?**

   Avoid cracking the flooring or joists, and avoid shooting through the floor.

11. **What information is contained in the site drawings?**

    All data pertinent to the site and installation

CHAPTER 3

1. **To what standard must the AC power wiring be performed?**

   The National Electrical Code plus any other local ordinances

3. **What is the IR drop if resistance is 0.04 ohms, and current is 10 Amperes?**

   0.4 volts

5. **What color is used for pair six in a twenty-five pair cable?**

   Red/Blue

7. **What is a "shiner" and how long is acceptable?**

   A shiner is a wire that has been stripped too far. 1/8 inch is the maximum allowed.

9. **What is the number one safety concern about a Fiber-Optic cable?**

   The thing you should always be aware of when working with Fiber-Optic cable is that the cable can blind you if it's energized, and you can't tell by just looking at a cable whether it's energized or not.

CHAPTER 4

1. **What does NEC stand for?**

   National Electrical Code

3. **Name at least four items that use 120 VAC power at a site.**
   - Lighting
   - Tower lights
   - Air conditioning
   - Heating
   - Peripheral equipment
   - Outlets around the building

5. **What does a transfer switch do?**

   Switches between the generator and the line power

7. **What does "8's" mean when you are talking about florescent lamps?**

   The wattage is in multiples of 8 as opposed to 10's. You cannot mix lights and fixtures.

9. **What are the two operating modes for tower lights?**

   Daytime and nighttime

11. **What does an outlet tester do?**

    Confirms that all of the leads are wired correctly on an electrical outlet

APPENDIX A: ANSWERS TO ODD-NUMBERED QUESTIONS

CHAPTER 5

1. **What are the components of a 24 VDC or 48 VDC power system?**

   The complete system includes the batteries, charger, power distribution panel, fuse panels, cabling, and alarm system.

3. **How often should gel-cell batteries be replaced?**

   Every five years

5. **Describe how an indicator fuse works.**

   When the fuse blows, the power is conducted to the alarm bus and the load is removed from the power source.

7. **What should the maximum voltage drop be on a given circuit?**

   1.0 volt

9. **What documentation is required for a 24 VDC or 48 VDC power system?**

   The identification of each breaker, and the current rating size of each fuse

CHAPTER 6

1. **What does MPE mean?**

   Maximum Permissible Expose, a level of the maximum amount of RF radiation that a person should be exposed to.

3. **What is ingress?**

   Other radio signals getting into an RF cable or system

5. **How is the loss of a cable usually expressed?**

   dB lost per 100 feet

7. **Why are jumpers used?**

   Larger cables cannot be bent or do not fit.

9. **How can cables be tested to confirm that they are made properly?**
   - Continuity testing with a DMM or VOM
   - Testing with a wattmeter

- Testing with a Time Domain Reflectometer (TDR)
- Testing with a Frequency Domain Reflectometer (FDR)

11. **What is the best method to check an antenna installation?**

    A Frequency Domain Reflectometer sweep is the best way to check an antenna installation.

CHAPTER 7

1. **What component has the greatest effect on the range of a radio system?**

    The antenna

3. **What is the most important part of an antenna or tower installation?**

    Safety!

5. **What is the gain of a 4 element omni-directional exposed dipole antenna?**

    6 dB

7. **What device is used as an aid to pull a coax up a tower?**

    A cable hoisting grip

9. **What is capillary action?**

    A property of physics that allows moisture to creep into tight spaces

11. **What is the device called that pumps air into an air-filled transmission line?**

    A dehydrator

CHAPTER 8

1. **What are the leads called in a telephone circuit?**

    Tip, ring

3. **What side of a punch down tool has the cutting edge?**

    The yellow side

APPENDIX A: ANSWERS TO ODD-NUMBERED QUESTIONS

5. **What does MDF mean?**

   Main Distribution Frame

7. **What is the tip and ring pair direction in a private line?**

   Transmit audio

9. **What does a loopback set do?**

   Allows the telco to test and set the audio levels remotely

11. **How many voice channels can be carried by a standard T1 circuit?**

    Twenty-four

CHAPTER 9

1. **How many ground rods should be used around a tower?**

   At least four

3. **What gauge wire should be used for the outside ground ring?**

   #6 or better

5. **What size wire should be used for the inside ground ring?**

   #6

7. **Can you drill into the tower leg to secure the ground cable to the tower?**

   No, never!

9. **At how many different points from the tower to the building must the coaxial cable be grounded?**

   Three

CHAPTER 10

1. **Name the four access points that must be protected?**

   - AC power
   - Telephone lines
   - Coaxial cables
   - Transmission line center conductor

3. **What does a site's lightning protection protect against?**

   Near hits and tributary fingers

5. **What is the best type of antenna to use in areas that are extremely lightning-prone?**

   An exposed element grounded dipole antenna is immune to all but major hits of lightning.

7. **What are the three types of RF lightning protectors?**
   - Spark gap
   - Fuse
   - Shorted stub

9. **How does a fuse type lightning protector work?**

   The fuse type of lightning arrester has a thin conductor to allow the RF to pass through. A lightning hit, however, with its large current will blow the fuse and open the RF equipment to the antenna. The lightning then has a path via the frame of the lightning protector, which is always connected to a good ground via a large cable.

CHAPTER 11

1. **What does N.O. mean?**

   Normally open

3. **Can an alarm be transmitted if a wire breaks on an N.O. sensor circuit?**

   No

5. **Will a broken wire be detectable on an N.C. alarm circuit?**

   Yes

7. **What does nm stand for?**

   Nano-meter

9. **Name three types of Fiber-Optic connectors.**

   ST, FC, and SC

11. **What should be done before installing any Fiber-Optic cable?**

    Clean the ends—*always*.

APPENDIX B

Glossary

2G The current version of cellular and PCS telephones are called Second Generation, or 2G, mobile telephones.

2.5G The new mobile telephones that also allow limited internet access are called 2.5G phones.

3G The next generation of mobile and portable telephones that act like computers and telephones are called 3G (Third Generation) phones. There is no standard yet for what 3G is supposed to be.

802.11B A protocol for wireless data interconnection.

AC Generator A gas- or liquid fuel-powered engine that produces electricity.

ADA The Americans with Disabilities Act. A law passed by the United States Congress that assures that people with handicaps have access to all buildings and places.

Anchor A bolt that is attached to the floor or ceiling to keep a piece of equipment in place.

Attenuation The reduction or loss of signal as it passes through a device or cable.

Auto Start A mode for generators that allows the generator to automatically start whenever the AC power fails.

Battery The hot side of the power in a 24 VDC or 48 VDC system.

Bluetooth™ A protocol for wireless data interconnection.

Breaker A device in the load center that can be used to switch the power on or off. Breakers also trip off when the AC power current exceeds the rating of the circuit breaker.

Bus A large surface conductor that ties many components together, including the ground lead.

Capillary Action A tendency for molecules, such as those of water, to migrate into very small openings, such as the cracks or pores of electrical tape.

CDMA Code Division Multiple Access. A mode of cellular and PCS radio that allows multiple conversations in the same frequency spectrum.

Coaxial Cable A wire that consists of an outer shield and a center conductor.

Color The ring lead of a pair in telephone switchboard cable. The choices are normally blue, orange, green, brown, or slate.

Compression Lug A termination lead that uses mechanical compression to hold the wire or cable to the lug.

Conduit A metal or plastic pipe that is used to enclose electrical wires or Fiber-Optic cables.

CRT Cathode Ray Tube. The video monitor on a computer system.

Data Circuit An interconnection line that carries data between two points.

DB25 The normal connector for serial data or parallel data to or from a computer.

DC Blocking Connectors and adapters that will pass RF signals but will not allow DC power to pass.

Decibels A unit that relates one voltage or power level to another.

Dehydrator An air pump that pressurizes coaxial cable, using air for the insulation between the center conductor and the shield.

Demarc Abbreviation for Demarcation Point. The point where the telephone company's responsibilty ends and the telephone user's starts.

Dialtone The signal from the telephone company that indicates that the Central Office is ready to accept digits from the user of the telphone line.

Dipole The common length of an antenna. It consists of the hot side and the shield side, with a length of 1/4 wavelength each.

Discone A wideband antenna used by the military, airports, and ham radio operators.

DMM Digital Multimeter. A multipurpose meter to read voltage, current, resistance, and other parameters in electrical and electronic circuits.

Drip Hole A small hole at the bottom of an antenna element that enables moisture to escape.

Egress Radio signals that escape from a sealed coaxial system.

EIA Electronics Industries Association. An industry trade group that promotes the electronics manufacturing business in the United States.

FAA Federal Aviation Administration. The agency of the United States government that oversees all aspects of flying and airplanes. Since radio towers can interfere with airplanes, the FAA has jurisdiction over all towers over 200 feet high.

FCC Federal Communications Commission. The agency of the United States government that oversees all aspects of radio and wireless signals.

Ferro-Resonant A type of transformer that has special windings to counteract low and high voltage spikes and surges.

Fiber-Optic Cable A cable used to carry high speed Fiber-Optic signals.

Float The process of keeping the battery voltage up to the maximum level permitted.

Frequency Domain Reflectometry (FDR) A method of checking antennas and transmission lines using frequency pulses to look at the equipment.

Fuse A device that protects equipment from drawing too much current.

Galvanized A method where tin is bonded to steel to prevent rust.

GPS Global Positioning Satellite. A system of satellites that provides precise time and location measurements.

Grasshopper Fuse A fuse that trips an alarm when the fuse wire has been blown from too much current. Also called an indicator fuse.

Ground The common lead in a system; also the central point of a return system in a building.

Ground Ring A heavy conductor cable that encircles a room, building, or site. All components within the ring are tied to it.

GSM Global System for Mobile Communications. A popular communications protocol that is used around the world, including the United States.

Halo Ground A ground ring that circles the inside of a radio room, near the ceiling.

Halon A material used in the past thirty years to suppress fires in a closed area. It has been replaced in the last few years by FE36 because of Halon's toxicity to humans.

Harness There are actually two separate definitions for harness that are used in the radio world. The first is a rigging worn by tower climbers that is safer than a climbing belt. The second is an interconnecting wiring arrangement that is used to connect multiple antennas together to increase the gain of the antenna.

IEEE The Institute of Electrical and Electronic Engineers. A group of graduate electrical engineers that represents the industry and the membership of the group.

Ingress Radio signals that leak into a coaxial cable.

IR Infrared. The spectrum of light just below red that is commonly used in Fiber-Optic communications systems.

ISDN Integrated Services Digital Network. A system that combines data and voice on the same telephone line. ISDN has become obsolete with the advent of Digital Subscriber Lines (DSL), which operates four times faster than ISDN, and up to thirty times faster than a 28.8 K modem.

Jumper A small cable to interconnect components.

LAN Local Area Network. A computer integrated system within a building.

Leakage A phenomenon where radio signals can escape or enter a coaxial cable system because there is not 100% shielding in the system.

Lightning A phenomenon of nature whereby static electricity builds up to millions of volts, and then discharges to the ground or to other clouds. The frequently explosive discharge is called lightning.

Line The control leads that leave the equipment and go to the telephone company or to the off-premises equipment.

Line Frequency The 60 Hz frequency of power lines and generators.

Load Center A breaker box or fuse box for AC electrical power distribution.

Loopback A device that the telephone company uses to test its lines without having to visit a site.

Loss A degradation of a signal due to any factor or combination of factors.

Lumens A measurement level of light.

MAN Metropolitan Area Network. A system that covers an entire metropolitan area using wireless technologies.

MDF Main Distribution Frame. The wire interconnection frame where the wiring of a site or Central Office all ties together.

Megger A meter used to measure ground effectiveness.

MGP Main Ground Point. The central ground for a radio system and tower.

MPE Maximum Permissible Exposure. The maximum safe level of radio signal that a person can be exposed to when in close proximity to the transmitters or antennas.

Multimode A Fiber-Optic cable that is 62.5 um in diameter.

NEC National Electrical Code. A set of wiring standards that is agreed upon by the National Fire Protection Association, the Institute of Electrical and Electronic Engineers, and the International Association of Fire Chiefs. All electrical wiring must follow the standards of the NEC or better, depending upon the local building and wiring codes.

Neutral The return path on an AC circuit.

nm Nanometer. A unit of measure; one nanometer is one billionth of a meter.

NO-OX A compound used in compression lugs to prevent corrosion and oxidation.

Normally Closed (N.C.) A type of alarm system trigger where the contact of a switch or relay is normally making contact to complete a circuit. Breaking the circuit triggers the alarm.

Normally Open (N.O.) A type of alarm system trigger where the contact of a switch or relay only makes contact when the switch or relay is operated. Closing the contact triggers the alarm.

Ohms The unit of resistance. It is also the result of the voltage divided by the current.

PAN Personal Area Network. A network that is wireless and covers only a few feet in radius.

PCS Personal Communications System. A wireless radio system used for voice and data, which occupies the band from 1800 MHz to 2000 MHz.

PDA Personal Digital Assistant. A handheld device that is a calculator, database organizer, and in some cases, a computer terminal.

Plenum A space between a false ceiling and the actual top of the room.

Plenum Rating A standard that ensures that the wiring in the space above the ceiling does not give off toxic smoke in the event of a fire. This standard is found in hospitals and most new buildings.

Plumb A term used in the installation and construction industry meaning that the sides of a piece of installed equipment are straight, that vertical and horizontal supports line up, and that all equipment lines are set to match all other equipment lines.

Polarization The direction—horizontal, vertical, or circular—that a radio wave is emitted.

Polarized Plug An AC outlet plug that has one prong slightly larger than the other so that it can only fit into a polarized socket in one direction.

POTS Plain Old Telephone Service. The standard telephone service like you find at your home.

Private Line A telephone line that ties two or more dedicated locations together.

PSTN Public Switched Telephone Network. The normal switched service of the telephone network.

R66B A standard punchdown block that is used to terminate twenty-five pair telephone cables.

RFID Radio Frequency Identification. Small tags, similar to inventory control tags at department stores, that emit radio signals which can be detected by strategically located receivers, indicating the location of the RFID tag.

Ribbon Cable A cable that has all of the leads next to each other, so that it looks like a flat ribbon.

Ring The side of a telephone line that has the ringing voltage on it.

RJ11 The standard 4 pin or 6 pin jack that is used to connect a telephone station to the line.

RJ21 A twenty-five pair cable termination that has an Amphenol connector tied to one or both sides.

RJ45 The standard 8-conductor connector used in Private Line data circuits.

RS232-C A standard that defines voltages and lead designations in serial data circuits.

Shielding Any of several means to electrically isolate one circuit from another.

Shorted Stub A phenomenon where a short looks like an open 1/4 wavelength away, but at all other frequencies, it looks like a short. This is a very effective lightning protection device, because it shorts out any lightning that gets into a transmission line.

Single-Mode A Fiber-Optic cable that is 9 um in diameter. This prevents the signal from bouncing within the cable.

Spark Gap A type of lightning protector that is used on telephone lines, AC lines, and RF transmission lines.

Split Bolt A bolt that has a hollow center that is cut out, enabling a heavy ground lead to be secured to a ground ring.

Station The common term used to describe a telephone set.

Station Wire The four conductor cable used by the telephone company to extend the stations to more convienient locations.

Surge Arrester A lightning protector that acts whenever the voltage goes higher than a preset limit.

T1 A multiplexing standard that allows twenty-four circuits to be encoded onto two pairs of wire.

TDMA Time Division Multiple Access. A scheme where the same radio frequency is shared by multiple data streams, but with a time slot allocation to keep the streams separate.

Telematics The use of automated devices to control other devices.

Tip The side of a telephone line that does not have the ringing voltage on it.

Tower A structure used to support one or more radio antennas.

Tracer The tip lead of a telephone switchboard cable that is colored in white, red, black, yellow, or violet.

Transfer Switch The electrical panel that controls whether the AC power comes from the commercial lines or the on-site generator.

UPS Un-interruptible Power Supply. A power supply that has batteries and converts the battery direct current voltage to normal 120 volts AC. In the event of a 120 VAC power failure, the UPS seamlessly provides backup power; therefore, when the 120 VAC is not present, the batteries internal to the UPS system are providing all of the power for the attached system(s).

Varistor A device that shorts out high voltage spikes.

V-Bolt A V-shaped bolt used to harness equipment to a tower that has either flat, angle, or round member structures.

VF Voice Frequency. The normal audio leads in a system.

VOM Volt-Ohm-Meter. A multimeter used to measure voltage, resistance, current, and other parameters in electrical and electronic circuits.

WAN Wide Area Network. A LAN that is larger than a single building.

Wireless All equipment that uses no leads to communicate.

WireWrap A method where leads are spun onto posts that allows many connections in a very small space.

Yagi A directional beam antenna designed for a rather narrow band of frequencies. Yagi antennas have gain over dipole antennas.

APPENDIX C

ADDITIONAL RESOURCES

There are quite a few books, companies, and web sites that can be accessed to further your knowledge in the world of wireless. Some of the books include:

- *Electronic Communications Systems, Second Edition*, by Roy Blake.
- *Introduction to Telecommunications*, by Anu A. Gokhale.
- *Wireless Technicians Handbook*, by Andrew Miceli.
- *Telecommunications Wiring, Third Edition*, by Clyde Herrick.
- *Wireless Communications Technology*, by Roy Blake.
- *Electrical Wiring Industrial, Eleventh Edition*, by Robert L. Smith and Stephen H. Herman.
- *Radio Amateurs Handbook*, by ARRL (www.arrl.org).
- *Electrical Grounding, Sixth Edition*, by Ronald P. O'Riley.
- *The National Electrical Code, 2002*, by the NFPA.

Some of the manufacturers who have equipment and corresponding installation standards include the following:

- Motorola
- M/A Com
- Lucent Technologies, Inc.
- E. F. Johnson

- Nortel
- Alcatel
- Onan
- Decibel Products
- Sinclair
- EMR, Inc.

Some of the manufacturers of test equipment used in the wireless industry include the following:

- Acterna
- Anritsu
- Agilent
- Bird
- IFR
- General Dynamics Decision Systems (Formerly Motorola Government Electronics Division/Test Equipment Division)
- Fluke
- Ramsey
- Tektronix

There are hundreds of other manufacturers out there, and omission from this list does *not* mean that a manufacturer should not to be considered or should not be looked into further for use at sites. Test equipment is required to insure that all of the components are working properly.

There are also thousands of web sites that can help you with the installation and maintenance of wireless sites. The FCC web site, at www.fcc.gov, is a great site, and very well organized. You should visit it to learn more about the federal rules and regulations that govern wireless sites.

INDEX

Symbols

1/4 wave dipole antenna 126
1/4 wave vertical antenna 126
100BaseT 56
10Base2 Ethernet 56
10BaseT 56
110-punch block 51
2 element exposed dipole antenna 131
3G 229
4 element exposed dipole antenna 132
8 element exposed dipole antenna 133

A

AC circuit 65, 68
AC generator 13
AC load center 13
AC power 3, 45, 199
AC power line 13, 196, 198
AC power meter 45
AC voltage 198
AC wiring 81
Acceptance Test Procedure 21
air conditioning system 9
alarm 15
alarm cable 45
alarm circuit path 94
alarm, fuse 94
alarm sensor 57
Alternating Current 66
Amphenol connector 48, 51, 170, 173
anchor 37
antenna 1, 124
antenna, 1/4 wave dipole 126
antenna, 1/4 wave vertical 126
antenna, beam 136, 138
antenna, dipole 126, 138
antenna, dish 142
antenna, exposed dipole 131
antenna, folded 1/4 wave 130
antenna, microwave dish 142
antenna, panel 144
antenna, sectorized 105, 144
antenna support 3
antenna transmission line 13
antenna, VHF/UHF 138
antenna, Yagi 136
attenuation 61, 206
azimuth 143

B

backup power 3, 5, 9, 65, 69
backup power system 83
bandwidth 130
battery 75, 87
battery backup system 199
battery float voltage 98
battery power 3
battery power system 3
battery string 88, 90
beam antenna 136, 138
beamwidth 142
block, RJ11 58
Bluetooth 232, 233
BNC connector 56
bracket mount 154
breaker 83, 93
breaker box 83
breaker rating 71
bridal ring 166
building ground ring 188
bus bar 72

C

cabinet system 7
cable, Alarm 45
cable bracket 154
cable, coaxial 56, 57, 103, 105, 117, 126, 152, 158, 185, 186, 205
cable connection 28
cable, data 51
cable, data communications interface 45
cable drawings 61
cable entrance 28
cable entrance plate 183
cable, Ethernet 56
cable, feeder 93
cable, Fiber-Optic 45, 58, 218, 222, 223
cable, ground 197
cable, peripheral 212
cable placement 23
cable, power 92
cable, RF 101, 103, 105
cable, RF coaxial 45
cable, RS-232C interface 51
cable, telco 50
cable, telephone 45
cable termination 45
cable tie 59
cable, voice frequency 45, 47
cabling, inter-bay 44
cadweld 191
cadwelding 188
calculated load 68
Category 3 51
Category 5 51, 56
Category 5 jack 51
CDMA 228
cellular 6, 228
central alarm controller 215
central ground post 183
Central Office 165, 173
Champ connector 48
charger 90
circuit breaker 71, 93, 97
circuit breaker box 45, 71
circuit breaker, thermal 72
coaxial cable 56, 57, 103, 105, 117, 126, 152, 158, 185, 186, 205
coaxial transmission line 13, 186
Code Division Multiple Access 228
computer interface 51
conduit 45
conduit entrance 29
connection, cable 28
connector 106
connector, Amphenol 48, 51, 170, 173
connector, BNC 56
connector, Champ 48
connector, DB9 55
connector, discrete 107
connector, interseries 107
connector, male 50
connector, nine-pin 55
connector, RF 151
connector, RJ-45 56
connector, RJ11 168, 173
connector, RJ45 173
connector strip 77
connector, twenty-five pair 47
continuity 96
corner reflector 138
CRT terminal 7
cycle 83

D

D Sub-Miniature series connectors 55
data cable 51
data circuit 212
data communications interface 51
data communications interface cable 45
data line 13
data terminal 51
DB25 56
DB9 connector 55
DC blocking adapter 106
DC power 45
decibel 129, 223
dedicated site 6
dehydrator 160
demarc block 165, 170
demarc point 165
demarcation point 165
digital multimeter 117
dipole antenna 126, 138
Direct Current 88
Discone 138
discrete connector 107
dish antenna 142
DMM 61, 82, 97, 117, 158, 217
documentation 97
drip hole 148
drip loop 156
dropout 83

E

Effective Radiated Power 129
egress 105

EIA 213
electrical noise 83
electromagnetic wave 136
Electronics Industries Association 213
elevation 142
equipment bay 30, 46
equipment drawing 40
equipment rack 26
ERP 129
Ethernet 51, 56, 228, 230
Ethernet cable 56
exposed dipole antenna 131
exposed element antenna 204
exposed element grounded dipole antenna 202
external cable 30
external wiring 45

F

FAA 78
FAA permit 18
FCC 78, 233
FCC license 18
FCC tower registration 18
FDR 61
FE36 19
Federal Aviation Administration 78
Federal Communications Commission 78, 228
feeder cable 93
female termination 50
ferro-resonant transformer 200
Fiber-Optic jumper 58
Fiber Optics 212, 217
Fiber-Optic cable 45, 58, 218, 222, 223
Fiber-Optic system 211
fiberglass antenna 204
fire protection 19
fire suppression system 19
float 87
float voltage 98
fluorescent light 12, 77
folded 1/4 wave antenna 130
frequency bandwidth 136, 138
frequency domain reflectometer 61, 117, 159
Frequency Domain Reflectometry 156
fuel cell 9
full current mode 90
fuse 71, 93, 97
fuse alarm 94
fuse, grasshopper 87, 94
fuse, indicating 94
fuse, indicator 93
fuse, main 94, 97
fuse panel 46, 97, 98
fuse, pilot 97
fuse, primary 93
fuse-type protector 117

G

gap arrester 206
gas tube protector 117
gel electrolyte 90
gel electrolyte batteries 90
gel-cell batteries 90
gel-cells 90
generator 9, 69
Global Positioning Satellite receiver 146
GPS 105, 146
GPS antenna 105
GPS receiver 146
grasshopper fuse 87, 94
ground 65, 87
ground bus 185
ground cable 197
ground conductor 78
ground continuity tester 193
ground, isolated 34
ground lead 77
ground ring 181, 185
ground ring, building 188
ground ring, tower 188
ground rod 13, 188
ground system 3
grounding 13
grounding cable 28
grounding kit 155, 186

H

halo ground 181, 188
Halon 19
handshaking 55
hoisting grip 153
horizontal plane 136, 143
horizontally polarized 142, 143
hot lead 82

I

impedance 105, 156
indicating fuse 94
indicator fuse 93
induction 116

257

ingress 105
inner conductor 117
inside ground ring 188
inter-bay cable 30
inter-bay cabling 44
inter-bay wiring 44
interconnect block 51
interface, computer 51
interface, LAN 56
interseries connector 107
ISM 144
isolated ground 34
isotropic radiator 127

J

jack, Category 5 51
jack, RJ11 58
Journeyman Electrician 68
Jumper 116
jumper 120, 151, 223
jumper clip 50
jumper wire 169
junction box 75

L

LAN 51, 56, 230
LAN interface 56
layout drawing 40
lead-acid batteries 90
LED indicator 82
light, fluorescent 12
lightning protection 13, 80
lightning protector 106, 116, 197
lightning protector, RF 206
line driver 55
line loss 5
load center 45, 65, 69, 71,
    72, 75, 77, 83
load center connection 72
Local Area Network 51, 56, 230
log periodic 138
logarithm 128
loopback 173
loopback set 173, 176
low voltage circuit 216
lumen 77

M

main current draw 90
Main Distribution Frame 167

main fuse 94, 97
main power board 93
main radiating element 126
main transmission cable 151
main transmission line 154
male connector 50
MAN 228, 229
Maximum Permitted Exposure 103
MDF 167
megger 193
Metropolitan Area Network 228
microwave dish antenna 142
modem 7, 55
monitoring system 7
MPE 103

N

National Electrical Code 19, 66
National Fire Protection Agency 19
NEC 19, 67, 76
Network Operations Center 175
networking 227
neutral lead 77
neutral line 65
NFPA 19
nine-pin connector 55
NO-OX 87, 92
NOC 175
normally closed circuit 215
normally open circuit 215

O

Ohm's Law 91
operating frequency band 130
OSI Seven-Layer Model 228
outer conductor 117
outlet box 75
outlet tester 82
outside ground ring 188
overhead rack 30, 36

P

pad 1, 6
PAN 231
panel antenna 144
PCS 6, 144, 228
PCS telephone 227, 231
penthouse site 5
performance testing 61
peripheral 7

peripheral cable 212
peripheral data equipment 212
peripheral equipment 2, 4, 44, 56
peripheral outlet 45
Personal Area Network 231
Personal Communications Services 228
Personal Communications Systems 144
photocell 80
pilot fuse 97
PL circuit 163
Plain Old Telephone Service 166
plenum 21
plenum cable 21
plenum rating 21
polarity 96, 98
polarity, reversed 96
polarization 142
POTS 166
POTS line 172, 201
power, AC 3
power, backup 3, 5, 9, 65
power, battery 3
power board 46, 93, 97
power cable 92
power cutoff switch 71
power distribution board 98
power distribution panel 46, 93
power line 28
power line, AC 196, 198
power outlet 75
power system 3, 88
primary fuse 93
Private Line 201
private line circuit 163
protector, fuse-type 117
protector, gas tube 117
PSTN 172
Public Switched Telephone Network 172
punch block 50, 51, 169
punch down tool 166
punch tool 51
punchdown block 50

R

R66B 50
R66B block 166
raceway 58
rack 46
rack stringer 34
radio frequency 101
Radio Frequency Identification 234
radio frequency signal 57

radio site 1
radio system 1
raised floor 36
rating, Category 5 51
reflecting ground 126
remote connection 78
remote monitoring system 213
reversed polarity 96
RF cable 101, 103, 105
RF cable drawing 40
RF coaxial cable 45
RF connector 151
RF energy 103, 118
RF equipment 206
RF interference 51, 213
RF level 16
RF lightning protector 206
RF signal 206
RF system 101
RFID 234
ring ground 183
RJ-45 connector 56
RJ11 block 58
RJ11 connector 168, 173
RJ11 jack 58
RJ14 168
RJ21 50
RJ21 block 165, 166
RJ21 or Amphenol™ connector 173
RJ45 connector 173
RS-232C 51, 57
RS-232C interface cable 51
RS232 213

S

sectorized antenna 105, 144
short circuit 88
shorted quarter wave stub 117
shorted stub 206
side stringer 34
signal attenuation 223
signal range 124
sine wave 83
Single Pole–Double Throw switch 77
site drawing 40
site manager 27, 34, 76
SMR 144
solar power 9
solar power panel 9
spark gap 206
SPDT 77
Specialized Mobile Radio 144

259

spike 83
spindle 166
static electricity 130
string, gel-cell 90
stud gun 38
surge 83, 198
surge arrester 199
surge suppressor 200
system, battery power 3
system engineer 34

T

T1 connection 175
T1 line 173
T1 telephone line 229
TCP/IP 228, 230
TDMA 228
telco cable 50
Telematics 227
telephone board 166, 167
Telephone cable 45
telephone line 28
telephone station 57, 168
telephone wiring board 163
termination, female 50
thermal circuit breaker 72
thermal sensing 71
Third Generation Personal Communication Services 229
Time Division Multiple Access 228
time domain reflectometer 117, 158
tip and ring 173
tower ground ring 188
tower lighting 65
transfer switch 70
transmission line test set 175
trickle charge mode 90
twenty-five pair connector 47, 166

U

UL 19
Underwriter's Laboratories 19
Uninterruptible Power Supply 73
UPS 13, 73, 90, 199
UPS system 73, 83
user installable block 51

V

Vector Network Analyzer 61, 159
vertical plane 136
vertical stacking connection 51
vertically polarized 142, 143
VHF/UHF antenna 138
VNA 61
Voice Frequency cable 45
voice frequency cable 47
volt-ohmmeter 117
voltage, AC 198
voltage drop 97
voltage, float 98
voltage spike 116, 198
voltmeter 96
VOM 61, 117

W

WAN 230
wattmeter 117
wavelength 83
WECA 231
weight loading 38
Wide Area Network 230, 232
Wireless Ethernet Compatibility Alliance 231
Wireless LAN 230
Wireless Local Area Network 230
wireless site 1, 66
wireless system 124
WireWrap terminal block 51
wiring conductor 75
wiring, external 45
wiring harness 150
wiring, inter-bay 44

Y

Yagi antenna 136